SpringerBriefs in Applied Sciences and Technology

For further volumes:
http://www.springer.com/series/8884

Karl-Heinz Schwalbe · Ingo Scheider
Alfred Cornec

Guidelines for Applying Cohesive Models to the Damage Behaviour of Engineering Materials and Structures

 Springer

Karl-Heinz Schwalbe
ehem. GKSS-Forschungszentrum
Geesthacht
Geesthacht
Germany

Alfred Cornec
Helmholtz-Zentrum Geesthacht
Geesthacht
Germany

Ingo Scheider
Helmholtz-Zentrum Geesthacht
Geesthacht
Germany

ISSN 2191-530X   ISSN 2191-5318   (electronic)
ISBN 978-3-642-29493-8   ISBN 978-3-642-29494-5   (eBook)
DOI 10.1007/978-3-642-29494-5
Springer Heidelberg New York Dordrecht London

Library of Congress Control Number: 2012937865

Printed on acid-free paper

Springer is part of Springer Science+Business Media (www.springer.com)

Contents

Summary

This document describes the guidelines for the application of cohesive models to the evolution of damage in materials and structures. The material is characterised by a specific constitutive law, the cohesive law, which is assigned to cohesive elements. These cohesive elements are embedded in a finite element mesh of the structure under consideration.

The cohesive law relates the tractions acting on the cohesive element surface to the material separation and is therefore called traction-separation law (TSL). It can be described by a function, which includes three material parameters:

- The cohesive strength, T_0,
- The cohesive energy, Γ_0, and
- The critical separation, δ_0,

which are determined by hybrid experimental/numerical procedures. In addition, the shape of the function must be determined or predefined for a complete description of a cohesive law. If, however, for a given class of materials the shape of the cohesive law is fixed, then only two parameters are needed. In this procedure, the cohesive stress and the cohesive energy are considered the relevant parameters.

The methods described in this procedure can be used for the determination of damage in materials and structures with and without pre-existing flaws, and hence for structural assessment. They are based on experience with metallic materials gained at the Helmholtz-Zentrum Geesthacht (formerly GKSS) as well as taken from the literature. The first part of the suggested procedure describes the material characterisation. In the following section areas of application are compiled. A particularly interesting area of application is the prediction of slow, stable crack extension in thin-walled lightweight structures. Open issues are also briefly described.

In an appendix validation by means of test pieces and application to some structural configurations are given.

Definitions

Traction–Separation Law
A set of equations (also known as cohesive law) describing the relationship between a stress and a separation within the cohesive elements, from zero separation up to loss of coherence.

Ductile Tearing
Crack extension due to micro-void formation and coalescence.

Cohesive Strength
Maximum stress of the traction–separation law.

Cohesive Energy
Area under the curve describing the traction–separation law

Damage
Irreversible process that causes material degradation and finally causes failure. The kind of damage and thus the affected material volume depends on the micro-mechanisms of damage.

Separation
Phenomenological description of material damage and failure by assuming that two initially connected material points separate by some damage process.

Finite Element Method (FEM)
Numerical method based on the principle of virtual work to effectively calculate the material behaviour by dividing a structure into elements with simple displacement formulations.

Cohesive Element
The cohesive element is a special numerical formulation within the framework of finite elements which is able to model the material separation by a cohesive law.

Global Mode I Fracture

Global loading which produces a typical mode I stress field.

Flat Fracture

Fracture surface which evolves normal to the applied loading direction under →*Global Mode I* Fracture conditions.

Slant Fracture

Fracture surface which evolves frequently for thin-walled panels. It turns to 45° (out-of-plane) with respect to the global crack propagation direction.

Normal Separation

Expression used only in the context of cohesive elements. Normal separation is the opening mode of the cohesive elements normal to its interface orientation

Tangential Separation

Like → *Normal separation* only used in the context of cohesive elements. Opening mode of the cohesive elements in the plane of the interface element.

Stable Crack Extension

Crack extension, which, under displacement control, stops when the applied displacement is held constant.

Fracture Resistance

The resistance a material exhibits to stable or unstable crack extension, expressed in terms of K, δ_5, J or CTOA.

Nomenclature

Dimensions

a Crack length
a_0 Initial crack length prior to ductile crack extension
t Specimen thickness
W Specimen width

Tensile Properties

E Young's modulus
v Poisson's ratio
$R_{p0.2}$ Yield strength equivalent to 0.2 % proof stress
R_m Tensile strength
σ_Y Yield strength, general

Forces, Stresses and Displacements

F Applied force
h Stress triaxiality, calculated by σ_m/σ_{eff}
v_{LL} Load-line displacement
σ Normal stress
σ_{eff} Effective stress
σ_m Mean stress, also called hydrostatic stress, $\sigma_m = \sigma_{ii}/3$

Fracture Parameters and Related Quantities

Δa Crack extension
J Fracture resistance in terms of the experimental equivalent of the J-integral
J_i Value of J at initiation of ductile tearing
Ψ Crack tip opening angle

δ_5 Crack opening displacement measured at the surface of a specimen or structural component at either side of the original crack tip over an initial gauge length of 5 mm

Cohesive Parameters

T Cohesive stress
T_0 Cohesive strength
Γ_0 Cohesive energy
δ Separation
δ_0 Critical separation
δ_n Separation normal to the fracture surface
δ_t Separation tangential to the fracture surface

The SI Units to be used in this Procedure are:

F Force, kN
σ, T Stress, MPa
δ_5 Displacement, mm
Δa Crack extension, mm
J Experimental equivalent of J-integral, MPa m

Acronyms

CMOD   Crack mouth opening displacement
CTOA   Crack tip opening angle
CTOD   Crack tip opening displacement
TSL   Traction—separation law
FEM   Finite Element Method

Chapter 1
Scope and Significance

1.1 Objective

This document provides guidance for the application of cohesive models to determine damage and fracture in materials and structural components. This can be done for configurations with or without a pre-existing crack. Although the present document addresses structural behaviour, the methods described herein may also be applied to any deformation induced material damage and failure, e.g. those occurring during manufacturing processes.

Guidance is provided for the following elements of an analysis:

- Formulation of the cohesive law,
- Determination of the cohesive parameters,
- Validation of the cohesive model and application to structural assessment.

According to the experience gained by the authors and their colleagues at the Helmholtz-Zentrum Geesthacht, Institute of Materials Research, the methods described in this document are applicable to the behaviour of ductile metallic materials and structural components made thereof. Hints are also given for applying the cohesive model to other materials. Since experience in practical applications is limited to global mode I failure at this time, the main part of the document will be confined to this issue. However, the basis for the description of mixed mode failure will be given.

The model described here considers traction–separation laws implemented in interfaces, i.e. special elements inserted between facets of existing bulk finite elements. Other implementations of traction–separation laws inside continuum elements as a displacement jump are out of the scope of this document.

Due to its computer efficiency, the cohesive model is particularly suited for large amounts of crack extension, and hence for the assessment of thin-walled light weight structures.

K.-H. Schwalbe et al., *Guidelines for Applying Cohesive Models to the Damage Behaviour of Engineering Materials and Structures*, SpringerBriefs in Applied Sciences and Technology, DOI: 10.1007/978-3-642-29494-5_1, © The Author(s) 2013

1.2 Traction–Separation Law

Numerous formulations for a cohesive law have been developed. The choice of a specific formulation depends strongly on the class of material under consideration (e.g. ductile metals, concrete, polymers, etc.).

Within this Procedure, several traction–separation laws are discussed based on their typical application and on their ease of use, since some of them are already implemented in commercial codes. However, not all implemented laws are well suited for the problem under consideration. Guidance is given on the choice of the traction–separation law and also for the implementation of a new cohesive element with a specific law.

1.3 Determination of Cohesive Parameters

In this Procedure, the two parameters: cohesive strength, T_0, and cohesive energy, Γ_0, are used to describe the cohesive law of a given shape quantitatively. These parameters are determined by means of hybrid methods combining experiments and numerical simulations. Methods for parameter optimisation are also given.

Chapter 2
Introduction

Due to the developing nature of the cohesive model, this document is supposed to provide guidance on practical application rather than a document written in strict procedural form. As potential users may be less familiar with numerical damage models than with classical fracture mechanics, a brief introduction of the cohesive model is deemed useful and will be provided in this chapter. For more in-depth studies see the fast growing literature in this area, e.g. the papers listed in the Bibliography section.

After the guidelines detailed in Chap. 3, areas of application are described in Chap. 4, some open issues are given in Chap. 5, showing the developing nature of the model and providing hints for further research. According to the experience gained at GKSS, the focus of the present document is on structural assessment, considering structural components made of bulk material and their welds. In an appendix worked examples are demonstrated.

2.1 Motivation for Applying Numerical Damage Models

Structural components containing crack-like flaws, or supposed to contain such flaws, are commonly assessed using the concepts of classical fracture mechanics. These concepts have become mature in the sense that both the characterisation of structural materials and assessment methods have already been cast into national and international standards, standard-like procedures, and codes. A comprehensive overview is given in [1]. However, the limits of classical fracture mechanics came into the focus of structural integrity research as higher exploitation of the mass of an engineering structure became of increasing importance due to the increasing awareness of the limited availability of raw materials and fossil fuel.

In the framework of fracture mechanics, cracked bodies are basically treated in a two-dimensional manner, notwithstanding the fact, that frequently

K.-H. Schwalbe et al., *Guidelines for Applying Cohesive Models to the Damage Behaviour of Engineering Materials and Structures*, SpringerBriefs in Applied Sciences and Technology, DOI: 10.1007/978-3-642-29494-5_2, © The Author(s) 2013

three-dimensional finite element analyses of test pieces and structural components are performed. The problem is that a fracture mechanics material property, either in terms of a single-valued parameter, commonly known as fracture toughness, or a relationship between the crack extension resistance and the amount of crack extension, is determined under circumstances describing the near-tip stress and strain fields under limiting conditions. These conditions are usually plane strain, more recently plane stress conditions have also been considered to account for the requirements of light-weight structures which are usually characterised by thin walled design [2], see also [3]. However, the conditions a structural component is under are frequently unknown and can substantially deviate from the test conditions for the determination of the fracture properties to be used for the assessment of the component. This problem is known as the transferability problem in classical fracture mechanics.

The advent of numerical damage models is providing a new approach to structural assessment in that these models deal with the damage events in the near-crack tip process zone which are embedded in the global finite element model of the component. This way the global FE model prescribes the loading conditions the component is under onto the damage zone. If the global FE model allows three-dimensional analyses and if the damage model parameters are given as functions of the triaxiality of the stress state, then the transferability problem is ideally solved. The damage models most commonly used can be partitioned into two groups: Models based on micromechanical processes of damage and phenomenological models.

2.1.1 Micromechanics Based Models

For ductile metals, this class of models is based on porous plasticity which describes the effects of nucleation, growth, and coalescence of voids on the load bearing capacity of a material volume as a function of the local stresses acting on that volume. A number of variants of these models have been suggested, see e.g. [4]. Of these, the Gurson–Tvergaard–Needleman model is the most widely used one.

In the case of brittle materials, another problem arises, which is the scatter of the input variables, such that only a failure probability can be given. Deterministic models as summarized here cannot capture this phenomenon and therefore this issue is outside the scope of this document. For further information, refer to [5, 6].

2.1.2 Phenomenological Models

In this group, the models do not depend on a specific failure mechanism and can therefore be used for arbitrary damage. However, the evolution law may indeed be applicable to a specific class of materials only. Within the group a further distinction can be made by the way of implementation: Models which have a damage law embedded in the continuum formulation are called continuum damage models, whereas the cohesive models do not describe material deformation but only separation. In the following, only cohesive models will be considered for the modelling of damage and failure of materials and structure.

The cohesive model employs a material model [7, 8], which is represented by a traction–separation law describing the loss of load bearing capacity of the material as a function of a separation, irrespective of the physical details of damage occurring in an actual material. Hence, it can be applied to both ductile and brittle damage and failure processes. The absence of mesh dependence—in contrast to porous plasticity models—makes the cohesive models very attractive. A length parameter is already included in the model, the critical separation δ_0.

Of numerical damage models, the phenomenological cohesive model is the most user friendly one and has a number of advantages. Among other items, numerical robustness, the mesh insensitivity and the need of only two model parameters—as opposed to e.g. up to nine parameters for the ordinary GTN model of porous metal plasticity—make the cohesive model a suitable candidate for practical application.

NOTE   It is recommended to read the article in Ref. [9] which contains a fundamental discussion on the nature of the terms "material parameter" and "model parameter".

2.2 The Cohesive Model

In this chapter, a brief overview of cohesive models and their properties is given. Unless otherwise indicated, the following overview is based on [10]. Cohesive models describe damage and fracture in a wide range of materials at various length and time scales. These materials include metals, polymers, ceramics, concrete, fibre reinforced materials, wood, rock, glass, and others.

As early as 1960, Dugdale [11] introduced a strip—yield model with the idea of a cohesive force preventing a crack from extending. The magnitude of this cohesive force is equal to the yield strength of the material, σ_Y; strain hardening is not considered, i.e. the material is supposed to behave in an elastic-ideally plastic manner. Since the local stress is limited by the yield strength of the material, the occurrence of a physically unrealistic singularity at the crack tip is avoided, Fig. 2.1. The result of this analysis is the length of the plastic zone ahead of a crack in an infinitely wide sheet subjected to a crack opening Mode I load; it is valid for small scale as well as wide spread yielding until the applied stress reaches the yield strength. Later, Goodier and Field [12] applied the Dugdale model to the

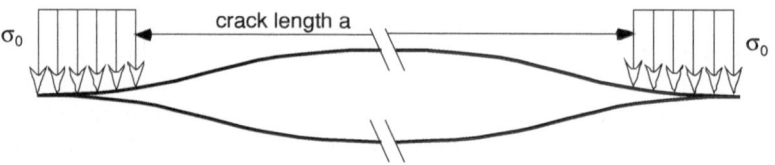

Fig. 2.1 The Dugdale model, after [11]

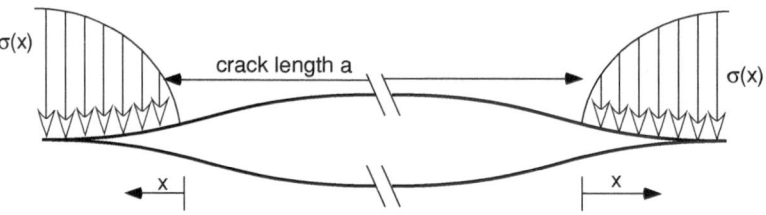

Fig. 2.2 The Barenblatt model, after [14]

determination of the opening profile of the crack including the crack tip opening displacement. Application of the model to a sheet of finite width is also reported [13].

The cohesive models in their present form date back to the work of Barenblatt [14] who replaced the yield strength with a cohesive law to model the decohesion of atomic lattices, Fig. 2.2. This way, the plastic zone was replaced by a process zone within which damage and fracture occur. The detailed processes are

- Plastic deformation;
- Initiation, growth, and coalescence of voids in ductile materials;
- Micro cracking in brittle materials.

Material degradation and separation are concentrated in a discrete plane, represented by cohesive elements which are embedded in the continuum elements representing the test piece or structural component. In both the Dugdale and Barenblatt models, the stresses along the ligament within the process zone do no longer depend on the applied load; they are now a material property. It should be noted that in Barenblatt's model the traction is expressed as a function of the distance from the crack tip, whereas the cohesive models actually in use define the traction as functions of the separation within the cohesive zone. Material degradation and separation are concentrated in a discrete plane, represented by cohesive elements which are embedded in the continuum elements representing the test piece or structural component.

To our knowledge, the first application of the cohesive model to the fracture behaviour of a material was performed by Hillerborg et al. as early as in 1976 [15], who used this model to describe the damage behaviour of concrete. This material has attracted much attention as to its characterisation using the cohesive model, see the work of the research groups of Bažant [16], Carpinteri [17], and Planas and

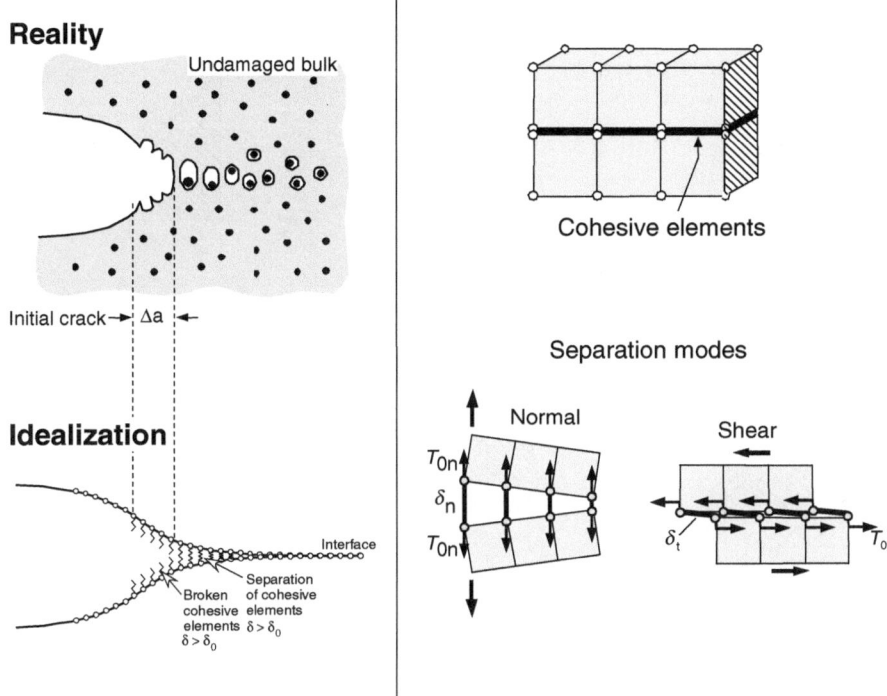

Fig. 2.3 Cohesive model: representation of the physical damage process by separation function within numerical interfaces of zero height—the cohesive elements

Elices [18]. For the other highly important class of engineering materials: metals and their alloys, pioneering work was performed by Needleman, Tvergaard, and Hutchinson. The first analysis of micro damage in ductile materials (particle debonding from a ductile matrix) was performed by Needleman in 1987 [19], and the first macroscopic crack extension in ductile materials was analysed by Tvergaard and Hutchinson [20]. Figure 2.3 shows how the physical process can be represented by the cohesive model. Experimental validation of the cohesive model for ductile materials has been investigated later on, e.g. by Yuan et al. [21]. Further details on applications of cohesive models are given in Chap. 4, for an extensive overview of the cohesive model see [7].

2.2.1 Traction–Separation Law

The constitutive behaviour of the cohesive model is formulated as a traction–separation law (TSL), which relates the traction, T, to the separation, δ, the latter

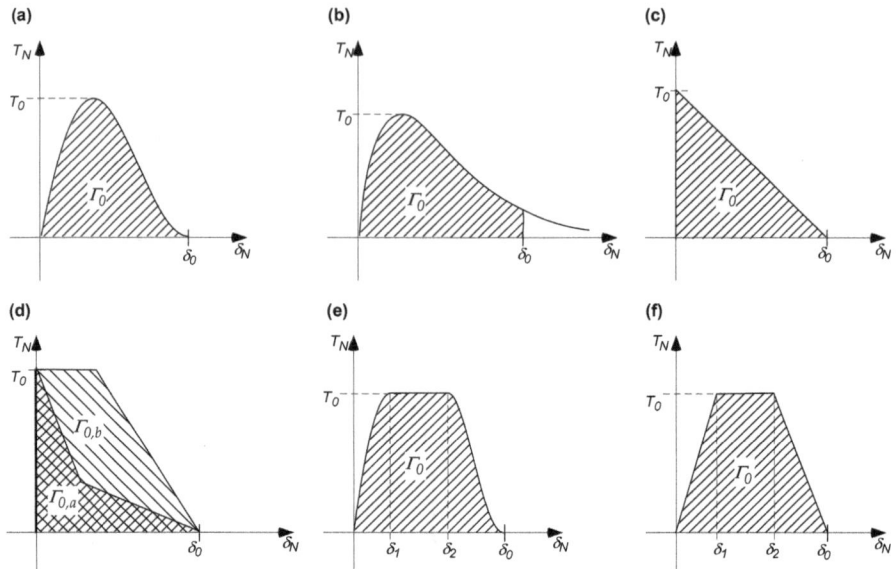

Fig. 2.4 Typical traction–separation laws: **a** Needleman [19], **b** Needleman [30], **c** Hillerborg [15], **d** Bazant [16], **e** Scheider [28], **f** Tvergaard and Hutchinson [20]

representing the displacement jump within the cohesive elements. A cohesive element fails when the separation attains a material specific critical value, δ_0. The related stress is then zero. The maximum stress reached in a TSL, the cohesive strength, T_0, is a further material parameter.

A host of traction–separation laws (TSL) have been suggested. (The term "cohesive law" is also being used instead of "traction–separation law"). Figure 2.4 gives an overview of frequently used shapes.

Brittle crack extension analyses of concrete were the first applications of the cohesive model. In purely brittle materials the traction–separation law can be easily identified, since all deformation that is inelastic can be assumed to be material separation. Therefore, the traction–separation behaviour can be determined from a simple uniaxial tensile test, in which the stress state is homogeneous and the elastic deformation can be subtracted from the global structural response. The resulting traction–separation law is often approximated by a linearly decreasing function, see Fig. 2.4c and [15, 22], etc. or by a bilinear function, Figure 2.4d, which has two additional parameters [16, 23].

For ductile metals, a TSL with a finite initial stiffness and a smooth shape as shown in Fig. 2.4a [19, 24], sometimes also with a softening curve approaching a horizontal asymptote and thus approaching zero traction at infinity (Fig. 2.4b), is often used in the literature. Other laws, which are more versatile by introducing additional shape parameters, have also been used.

For a given shape of the TSL, the two parameters, δ_0 and T_0, are sufficient for modelling the complete separation process. In practice, it has been proven useful

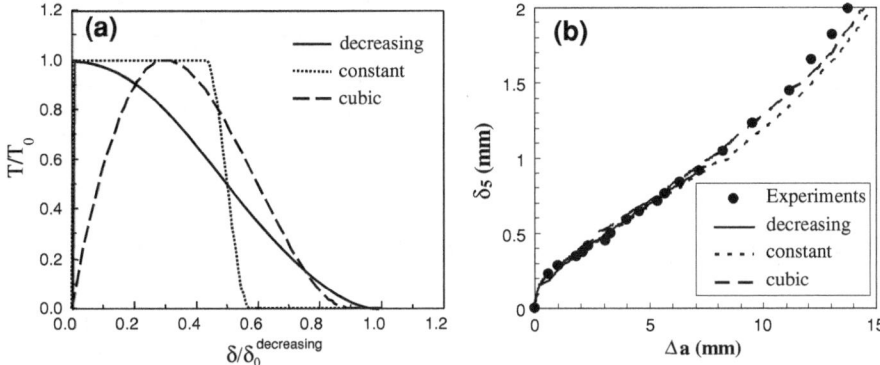

Fig. 2.5 **a** Three TSL's used for simulating a δ_5 R-curve; **b** δ_5 R-curve measured and simulated on a 50 mm wide C(T) specimen made of Al 5083 T321 [25]

to use the cohesive energy, Γ_0, instead of the critical separation. The cohesive energy is the work needed to create a unit area of fracture surface (in fact twice the unit fracture surface because of the two mating fracture surfaces) and is given by

$$\Gamma_0 = \int\limits_0^{\delta_0} T(\delta)d\delta \qquad (2.1)$$

There is a dispute as to whether the traction–separation law should have a finite slope right from the beginning, i.e. also small stresses lead to a material separation, or not. One can see that the shape used for ductile metals has a finite compliance in the beginning as shown in Fig. 2.4a and b, whereas for more brittle materials, Fig. 2.4c and d, the separation is set to zero until a the cohesive strength is reached. In order to avoid an unwanted "elastic" opening of the cohesive element, it is advantageous to have a high stiffness in the beginning. For example, in Fig. 2.4e and f the initial compliance can be defined based on additional shape parameters, which specify the separation at which the cohesive strength is reached.

If a traction–separation law is used, which starts without any separation until the cohesive strength is reached (Fig. 2.4c and d), a contact algorithm must be employed in the implementation of the cohesive element, since conventional elements can never have an infinite stiffness. Therefore, a TSL with a finite stiffness in the beginning is more convenient for implementation into an existing finite element code as a user element,

The choice of the TSL affects the magnitudes of the cohesive parameters, demonstrating the phenomenological nature of the cohesive model. This means that each traction—separation law requires a different set of parameters for a given problem as shown e.g. in [25]. Figure 2.5b shows an experimental δ_5 R-curve with cohesive simulations. It is seen that the three TSL's used, Fig. 2.5a, are able to models the experiment only with different sets of cohesive parameters, Table 2.1.

Table 2.1 Parameters
optimised for three TSL's to
simulate a δ_5 R-curve of the
aluminium alloy 5083 H321
[25]

	T_o (MPa)	δ_0 (mm)	Γ_0 (kJ/m^2)
Partly constant	560	0.024	10
Polynomial	590	0.043	14
Cubic decreasing	580	0.045	13

2.2.2 Tangential Separation and Mixed Mode Fracture

In the case of mixed mode loading, a tangential separation mode, usually desig-
nated Mode II and Mode III, accompanies the normally considered crack opening,
Mode I. In linear elastic fracture mechanics, a phase angle, Ψ_{LEFM}, can be defined

$$\Psi_{LEFM} = \arctan\left[\frac{K_{II}}{K_I}\right] \tag{2.2}$$

where K_I and K_{II} denote the stress intensity factors for crack opening Modes I and
II, respectively. In the context of the cohesive model, a tangential displacement, δ_t,
represents the additional shear mode and is superimposed to the displacement
normal to the crack plane (or plane of expected damage in the absence of a pre-
existing crack), δ_n. In analogy to Eq. (2.2), a phase angle for the cohesive model
reads [26]

$$\Psi = \arctan\left[\frac{\delta_t}{\delta_n}\right] \tag{2.3}$$

Alternatively, the phase angle can be expressed in terms of the energy release
rate corresponding to the cohesive energy by re-formulating Eq. (2.2)

$$\Psi = \arctan\sqrt{\frac{G_{II}}{G_I}} \tag{2.4}$$

In general the resulting displacement can be obtained from

$$\delta_{res} = \sqrt{\delta_n^2 + \delta_t^2} \tag{2.5}$$

It must be noted that there are almost as many mixed mode formulations for the
cohesive model as traction–separation laws. If the simple formulation, Eq. (2.5), is
used for an "effective" separation, the resulting traction in normal and tangential
directions is calculated by

$$\mathbf{T} = T(\delta_{\text{eff}})\left(\frac{\delta_N}{\delta}\mathbf{n} + \frac{\delta_t}{\delta}\mathbf{t}\right) \tag{2.6}$$

in which $\mathbf{n}$ and $\mathbf{s}$ are the normal and the tangential unit vectors of the cohesive
element, respectively. A similar formulation is given by an additional weighting

factor β for the tangential separation in Eq. (2.6) as introduced by [22], which then leads to

$$\mathbf{T} = T(\delta_{\mathrm{eff}})\left(\frac{\delta_N}{\delta}\mathbf{n} + \beta^2\frac{\delta_t}{\delta}\mathbf{t}\right) \tag{2.7}$$

Other formulations for the interaction of separation modes work without an effective separation, but with an explicit dependence on both the normal and the tangential separations, as e.g. developed by [27]:

$$T_N = T_0\,e\exp\left(-\frac{\delta_N}{\delta_0}\right)\left[\frac{\delta_N}{\delta_0}\exp\left(-\frac{\delta_T}{\delta_0}\right)^2 + (1-q)\left[1-\exp\left(-\frac{\delta_T}{\delta_0}\right)^2\right]\frac{\delta_N}{\delta_0}\right]$$
$$T_T = 2T_0\,e\,q\left(\frac{\delta_T}{\delta_0}\right)\left(1+\frac{\delta_N}{\delta_0}\right)\exp\left(-\frac{\delta_N}{\delta_0}\right)\exp\left(-\frac{\delta_T}{\delta_0}\right)^2 \tag{2.8}$$

An even more variable approach has been presented by [28], in which a multiplicative decomposition of the dependence on the separation components is used:

$$\begin{aligned} T_N &= T_{N,0}f(\delta_N)\,g(\delta_T) \\ T_T &= T_{T,0}f(\delta_T)\,g(\delta_N) \end{aligned} \tag{2.9}$$

In this case, the formulation for the shape of the TSL in the one-dimensional case, $f(\delta)$, is fully decoupled from the interaction function $g(\delta)$ and can be chosen independently.

It is worth noting that mixed mode loading causes path dependency, i.e. damage depends on how the two modes are activated during the loading process.

2.2.3 Cohesive Elements

This document considers cohesive elements as interface elements with two surfaces, which usually lie on top of each other in the undeformed state, i.e. they do not span a volume. If, however, the two surfaces have a finite distance in between, the resulting volume does not have any physical meaning, since it is not the strain in the element that is the relevant quantity, but the displacement jump between the surfaces. In order not to replace any real material by cohesive interface, the volume of the cohesive elements should be as small as possible and negligible compared to any other dimension in the model. In the framework of the finite element method, they have to be implemented corresponding to the surrounding continuum model, i.e. if the structure is modelled by 3D continuum elements, the cohesive elements must consist of surfaces as shown in Fig. 2.6a, if the structure is modelled in 2D or shell elements the cohesive elements reduce to line elements, see Fig. 2.6b and c. The difference between cohesive elements for plane strain/plane stress and shell structures is that the latter are defined in the three dimensional space. Therefore, any separation may be in-plane or out of plane, and the in-plane direction must be defined by the user, which can be done by a fifth node as shown in Fig. 2.6c.

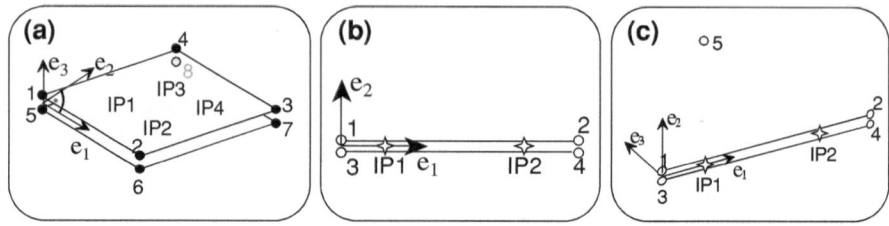

Fig. 2.6 Cohesive element library for 3D FE models (**a**), for plane stress/strain models (**b**), and for shell models (**c**)

Several commercial FEM solvers offer cohesive elements in their libraries. However, these elements are usually available for a subset of structural problems only, for example only for plane 2D and 3D problems, but not for shell meshes, etc.

2.2.4 Commercial Solutions for Cohesive Elements

The preferred method is to use those commercial codes such as ABAQUS, ANSYS, Marc, Zebulon and Warp 3D, which do already include cohesive elements. These elements are verified and tested, it is assumed that the simulation with these elements is robust and stable, and pre-processors can be used for their generation.

On the other hand, some features, which are being investigated in the scientific community, are not yet available in commercial codes. Usually, only a subset of element types can be used in combination with cohesive elements, and the number of possible traction–separation laws is limited as well. In many cases, a specific class of applications was intended when implementing the cohesive elements. To the authors' knowledge, no commercial code provides cohesive elements for shells.

Table 2.2 gives a short overview of some present implementations of cohesive elements in commercial finite element systems as well as the element types and traction separation laws available.

NOTE   For those users wishing to use their own elements or a finite element code without cohesive elements guidance is given in [29].

2.2.5 Crack Path

The cohesive elements embedded in the finite element model of the component to be assessed prescribe the path of crack extension. However, if the path a pre-existing crack is going to take is not known a priori cohesive elements have to be

Table 2.2 Commercial codes providing cohesive elements in their element libraries

Code	Cohesive elements for these structures	Traction–separation laws	Mixed mode	Remarks
ABAQUS	2D, 3D	Linearly decreasing, exponentially decreasing, tabular	Yes	Intended mainly for delamination
ANSYS	2D, 3D	Exponential (Needleman potential)	Yes	
MARC	2D, 3D	Bilinearly and exponentially decreasing function		Intended mainly for delamination
WARP3D	3D	Needleman potential	Yes	
Nastran	2D, 3D	Bilinearly, exponentially and exponentially decreasing function		Intended mainly for delamination
Zebulon	2D, 3D	Needleman potential	Yes	Incl. automatic contact activation after complete debonding
FRANC2D	2D	Linearly and bilinearly decreasing, exponentially decreasing		

placed anywhere the crack may chose its path, so that a specific array of cohesive elements becomes active and hence defines the crack path which reaches the failure conditions prior to the other arrays placed in the neighbourhood of the crack tip, Fig. 2.7. Consequently, it seems appropriate to equip every crack problem with as many possible crack paths as possible in order to avoid false predictions of the crack path. However, the amount of computational effort increases with the number of elements, and the choice of the crack path is not really arbitrary, since the most appropriate mesh is one consisting of triangles. Beside the large number of cohesive elements needed for such a simulation, the zigzagging path may cause numerical problems and lead to a mixed mode separation where in the real material a pure normal separation occurs. An additional problem arises when a TSL with an initial compliance is used. In this case, the total deformation even at small loads depends strongly on the compliance itself and the number of cohesive elements in loading direction, since all elements separate slightly in a homogeneous stress state.

2.2.6 Further Functionalities

For different classes of materials, a dependence of the cohesive parameters on various field quantities must be taken into account. Two examples are ductile

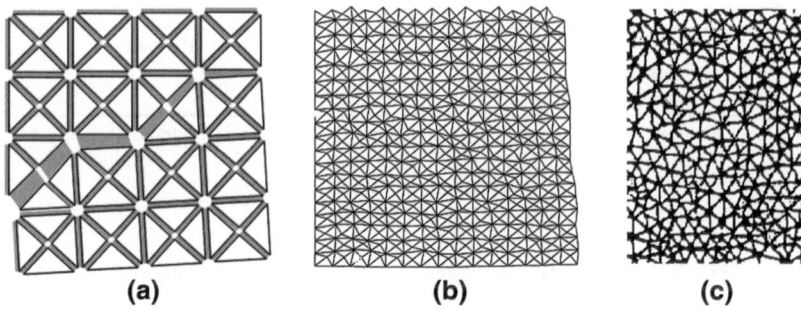

Fig. 2.7 Sketch showing meshes with cohesive elements between all boundaries of continuum elements; (**a**) is a regular mesh; (**b**) and (**c**) are irregular meshes. The mesh in (**b**) is derived by moving the nodes from mesh (**a**) randomly, whereas the nodes in mesh (**c**) are generated randomly

materials, where the cohesive parameters depend on the degree of triaxiality, or adhesives and plastics, for which the damage behaviour is strongly dependent on the local strain rate. Since cohesive elements do not account for either triaxiality or strain rate, these values must be imposed by the surrounding finite element model of the configuration considered. As in-depth studies are still lacking it is at present not possible to provide a systematic picture of these effects. Some hints will be given in Chap. 5.

References

1. Schwalbe, K.-H., Landes, J.D., Heerens, J.: Classical fracture mechanics methods. In: Schwalbe, K.-H. (ed.) Comprehensive Structural Integrity, Online Update, vol. 11. Elsevier, Oxford (2007)
2. Schwalbe, K.-H., Newman Jr, J., Shannon, J.: Fracture mechanics testing on specimens with low constraint—standardisation activities within ISO and ASTM. Eng. Fract. Mech. **72**, 557–576 (2005)
3. Heinimann, M., Zerbst, U., Dalle Donne, C. (eds.): Integrity of thin-walled light-weight structures—concepts and applications. Special issue of engineering fracture mechanics, vol. 76, pp. 1–164 (2009)
4. Pineau, A., Pardoen, T.: Failure of metals. In: Milne, I., Ritchie, R. O., Karihaloo, B. (eds.) Comprehensive Structural Integrity, pp. 684–797, chapter 2.06 (2007)
5. Riesch-Oppermann, H.: Fracture mechanics: probabilistic approaches. In: Encyclopedia of materials: science and technology, pp. 1–6 (2008)
6. Gutiérrez, M.A., de Borst, R.: Stochastic finite element methods. In: Comprehensive structural integrity, chapter 3.11, pp. 607–635 (2007)
7. Brocks, W., Cornec, A., Scheider, I.: Computational aspects of nonlinear fracture mechanics. In: Milne, I., Ritchie, R.O.K. (eds.) Comprehensive Structural Integrity: Fracture From Nano to Macro, pp. 127–209. Elsevier, Oxford (2003)
8. Elices, M., Guinea, G.V., Gómez, J., Planas, J.: The cohesive zone model: advantages, limitations and challenges. Eng. Fract. Mech. **69**, 137–163 (2002)

9. Brocks, W., Steglich, D.: Hybrid methods. In: Schwalbe, K.-H. (ed.) Comprehensive Structural Integrity, Online Update, vol. 11. Elsevier, Oxford (2007)
10. Cornec, A., Scheider, I., Schwalbe, K.-H.: On the practical application of the cohesive model. Eng. Fract. Mech. **70**, 1963–1987 (2003)
11. Dugdale, D.S.: Yielding of steel sheets containing slits. J. Mech. Phys. Solids. **8**, 100–104 (1960)
12. Goodier, J.N., Field, F.A.: Fracture of Solids. Wiley, New York (1963)
13. Schwalbe, K-H.: A modification of the COD concept and its tentative application to the residual strength of center cracked panels. In: Paris PC (ed.) Fracture Mechanics, Proceedings of the Twelfth National Symposium on Fracture Mechanics, ASTM STP 700, American Society for Testing and Materials, West Conshohocken (1980)
14. Barenblatt, G.I.: The mathematical theory of equilibrium cracks in brittle fracture. Adv. Appl. Mech. **7**, 55–129 (1962)
15. Hillerborg, A., Modeér, M., Petersson, P.E.: Analysis of crack formation and crack growth in concrete by means of fracture mechanics and finite elements. Cem. Concr. Res. **6**, 773–782 (1976)
16. Bažant, Z.P.: Concrete fracture models: testing and practice. Eng. Fract. Mech. **69**, 165–205 (2002)
17. Carpinteri, A.: Mechanical damage and crack growth in concrete. Martinus Nijhoff Kluwer, Dordrecht Boston (1986)
18. Planas, J., Elices, M., Guinea, G.V.: Cohesive cracks versus nonlocal models: closing the cap. Int. J. Fract. **63**, 173–187 (1993)
19. Needleman, A.: A continuum model for void nucleation by inclusion debonding. J. Appl. Mech. **54**, 525–531 (1987)
20. Tvergaard, V., Hutchinson, J.W.: The relation between crack growth resistance and fracture process parameters in elastic-plastic solids. J. Mech. Phys. Solids **40**, 1377–1397 (1992)
21. Yuan, H., Lin, G., Cornec, A.: Verification of a cohesive zone model for ductile fracture. J. Eng. Mater. Technol. **118**, 192–200 (1996)
22. Camacho, G.T., Ortiz, M.: Computational modelling of impact damage in brittle materials. Int. J. Solids Struct. **33**, 2899–2938 (1996)
23. Guinea, G., Planas, J., Elices, M.: A general bilinear fitting for the softening curve of concrete. Mater. Struct. **27**, 99–105 (1994)
24. Chaboche, J., Girard, R., Levasseur, P.: On the interface debonding models. Int. J. Damage Mech. **6**, 220–257 (1997)
25. Scheider, I., Brocks, W.: Effect of the cohesive law and triaxiality dependence of cohesive parameters in ductile tearing. In: Gdoutos (ed.) Proceedings of the XVI European Conference on Fracture, Alexandropolis, Greece (2006)
26. Walter, R., Olesen, J.F.: Cohesive mixed mode fracture modelling and experiments. subm. to Eng. Fract. Mech. **75**, 5163–5176 (2008)
27. Xu, X.-P., Needleman, A.: Void nucleation by inclusion debonding in a crystal matrix. Model. Simul. Sci. Eng. **1**, 111–132 (1993)
28. Scheider, I., Brocks, W.: Simulation of cup-cone fracture using the cohesive model. Eng. Fract. Mech. **70**, 1943–1962 (2003)
29. Scheider, I.: The cohesive model—foundations and implementation. In: GKSS Internal Report GKSS/WMS/06/02, (2002)
30. Needleman, A.: An analysis of decohesion along an imperfect interface. Int. J. Fract. **42**, 21–40 (1990)

Chapter 3
Material Characterisation

3.1 Fundamentals and Current Restrictions

The cohesive model can be used for any failure mechanism, i.e. it is applicable to any material and for any fracture mode, i.e. applicable to any loading. However, Mode II, Mode III and mixed mode fractures are of lower priority for engineering applications. In addition, no reliable procedure for parameter identification is available for any mode other than Mode I fracture as to the authors' knowledge.

Even in global Mode I fracture, additional fracture modes may occur locally. Namely this is the case for thin-walled structures, where a crack often turns such that the normal of the fracture surface is inclined by 45° to the main loading direction. The stresses acting on this plane are both of the normal Mode I and the shear Mode III types.

NOTE   If the normal of the fracture surface is parallel to the main loading direction, the local fracture mode is called *flat fracture*. If the fracture surface is inclined, the local mode is called *slant fracture*

The slant failure mode is described in Sect. 4.2. A few words about the parameter identification for slant fracture are given in Sect. 3.4.

In addition, at the current state, it is assumed that the crack path is known in advance, and therefore, only a single layer of cohesive elements is to be placed in the finite element model.

3.2 Traction–Separation Laws for Global Mode I Fracture

In this Procedure, the main focus is on numerical treatment of material separation under global Mode I conditions. Since all cohesive model quantities are given for the normal direction in this section, they are written without the subscript N.

K.-H. Schwalbe et al., *Guidelines for Applying Cohesive Models to the Damage Behaviour*   17
of Engineering Materials and Structures, SpringerBriefs in Applied Sciences and
Technology, DOI: 10.1007/978-3-642-29494-5_3, © The Author(s) 2013

Fig. 3.1 Traction-separation
law for ductile materials,
from [6]

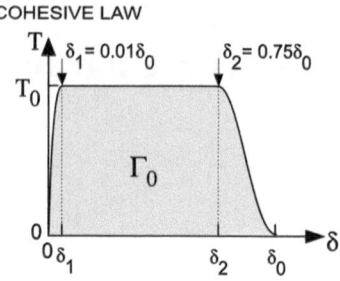

Although all classes of materials can be treated with the cohesive model, in this
document only ductile metals are covered, as already mentioned. The underlying
mechanism for this failure is void nucleation, growth and coalescence. The
treatment of brittle failure is outlined in Appendix 2.

The recommended TSL depicted in Fig. 3.1 is given by the following set of
equations:

$$T(\delta) = T_0 \begin{cases} \left[2\frac{\delta}{\delta_1} - \left(\frac{\delta}{\delta_1}\right)^2\right] & \text{for} \quad \delta < \delta_1 \\ 1 & \text{for} \quad \delta_1 < \delta < \delta_2 \\ \left[2\left(\frac{\delta-\delta_2}{\delta_0-\delta_2}\right)^3 - 3\left(\frac{\delta-\delta_2}{\delta_0-\delta_2}\right)^2 + 1\right] & \text{for} \quad \delta_2 < \delta < \delta_0 \end{cases} \tag{3.1}$$

The parameters δ_1 and δ_2 are set to $\delta_1 = 0.01\,\delta_0$ and $\delta_2 = 0.75\delta_0$.

The TSL is unequivocally determined by any two out of the following three
parameters

- Cohesive stress, T_0,
- Cohesive energy, Γ_0,
- Critical separation, δ_0.

According to Eq. (3.1), the area under the curve in Fig. 3.1 is given by

$$\Gamma_0 = T_0\,\delta_0\left(\frac{1}{2} - \frac{1}{3}\frac{\delta_1}{\delta_0} + \frac{1}{2}\frac{\delta_2}{\delta_0}\right) \tag{3.2}$$

Consequently, two cohesive parameters are sufficient to describe the TSL, of
which T_0 and Γ_0 have been chosen for this procedure.

NOTE   This traction–separation law is not implemented in commercial codes.
Therefore, if one wants to use commercial software only, the shapes that
are available there have to be used. More information on the question,
which shapes are available in which software, is given in Sect. 2.2.4.

NOTE   In several commercial codes, an exponential function as given by Eq.
(2.8) and shown in Fig. 2.4b, or a cubic polynomial function (Fig. 2.4a)
can be used. Even though the stress plateau is not very pronounced, these
types are often used in the literature for ductile fracture.

NOTE   The values of the cohesive parameters are a function of the TSL chosen.
        They have different magnitudes for different TSL's, see Fig. 2.5.

If the traction–separation law chosen is given by tabular input of data points of
the curve, the following procedure is recommended:

- The initial slope of the TSL should be as steep as possible. As a rule of thumb,
  the cohesive strength, T_0, should be reached before $0.05\delta_0$.
- A constant stress part should terminate at $\delta \leq 0.75\delta_0$, then the cohesive stress
  should decrease to zero at δ_0.
- If possible the corners of this multi-linear representation should be rounded by
  additional points.

NOTE   If the user is able to implement a contact algorithm which allows having
        an infinite stiffness up to the cohesive strength, this can also be used.

3.3 Finite Element Simulations with Cohesive Elements

3.3.1 Mesh Generation

In commercial finite element mesh generators it is usually not possible to define
interface elements without any initial volume as described in Sect. 2.2.3. Alternatively, the cohesive elements implemented in commercial codes can handle
interface elements with a finite volume in the undeformed state instead. Therefore,
the recommendation for mesh generation is as follows:

Define a thin layer of elements along the prospective crack path. The area of
one cohesive element (or the length of the line for 2D problems), in the following
called the characteristic size, w_{el}, must be chosen appropriately. The following
recommendations are given:

The height of the element layer h_{el} should meet the following requirements:

- $h_{el} \ll w_{el}$
- $h_{el} \ll \delta_0$
- $h_{el} \ll$ geometrical quantities (a, B, W, etc.)

NOTE   If the FE code requires a cohesive element with $h_{el} = 0$, the interface
        elements can be generated as stated before, and the nodes are shifted
        afterwards such that the top and bottom nodes coincide, which is possible
        in most of the FE pre-processors.

The cohesive element must be small enough to cover any stress gradient,
especially in the direction of separation.

In any case, the size must be small compared to the characteristic fracture
mechanics length scale, i.e. the radius of the plastic zone or the size of the HRR field.

As a rule of thumb, the size of the element in crack propagation direction
should be in the range of 50–250 μm for ductile materials. For 3D problems, the

Fig. 3.2 Basis for the
determination of material
separation in a cohesive
model framework

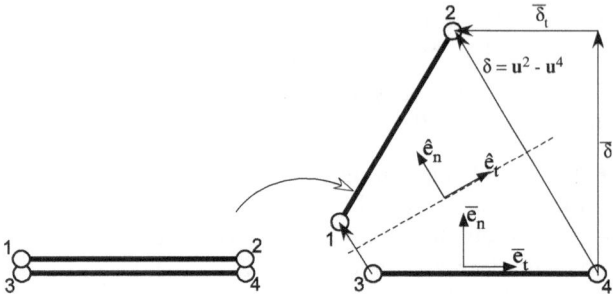

size in crack front direction must be such that it is able to cover any crack front
tunnelling. In addition, a high aspect ratio (say higher than 1:5) must be avoided.

3.3.2 Stress Calculation

Without going into the details of finite element fundamentals, it is important to
know about the residual nodal force calculation within a cohesive element. Two
aspects have to be considered:

1. The position of the reference plane, and
2. The calculation of the integration point area.

Re 1: A possible deformation of a 2D cohesive element according to Fig. 2.6 is
shown in Fig. 3.2. During this deformation, the upper surface has rotated by an
angle of 60° while the lower surface has kept its original position. In this case, the
distribution of the separation in a normal and tangential component depends on the
definition of the reference plane. If the original position, i.e. the lower surface
(coordinate system $\{\bar{e}_n, \bar{e}_t\}$ in Fig. 3.2), is used as reference for the calculation of
the normal and tangential separation, the separation vector between the nodes 2
and 4, $\boldsymbol{\delta} = \mathbf{u}^2 - \mathbf{u}^4$, has a normal and a tangential component. In contrast, the
material separation consists of normal separation only, if the centre between upper
and lower surface (the dashed line, coordinate system $\{\hat{e}_n, \hat{e}_t\}$ in Fig. 3.2) is used
as reference. In general, the latter should be used for any analyses.

Re 2: The calculation of the cohesive traction also deserves special attention.
The constitutive behaviour implemented in the finite element code can be written in
terms of nominal or true stresses as a function of separation. If nominal stresses are
used, the element stiffness matrix calculates the nodal forces based on the original
integration point area. However, in a crack propagation analysis of ductile materials
this area may change significantly due to plastic deformation of the surrounding
continuum elements. It may happen that due to plastic localization, the cohesive
strength is not reached at all in the cohesive element and hence no crack propagation
can be achieved.

NOTE   In commercial codes it must be checked whether nominal stresses (based on the original integration point area) or true stresses (based on the current integration point area) are used throughout the analyses. This can be performed either by a single element finite element calculation or by reading the manual (if available).

3.3.3 Numerical Convergence

Crack extension simulations of ductile materials are always strongly nonlinear. In most cases they have three different types of nonlinearity:

- geometrical nonlinearity,
- material nonlinearity due to plastic deformation,
- contact nonlinearity due to material separation.

Of course, these numerical issues may lead to convergence problems. Therefore, the rate of convergence should be improved by setting residual convergence controls explicitly. Since all commercial codes have their special settings, no specific recommendations can be given. However, in general the following issues should be considered:

- The number of increments in the simulation should be rather high in order to obtain an accurate solution.
- If the finite element program allows, Automatic Time Stepping should always be used.
- The maximum number of iterations within an increment as well as the number of cutbacks during automatic time stepping should be increased from the default. Full Newton iteration should be utilized if possible and an additional line search algorithm within an iteration is beneficial, too.
- If the displacement jump at a specific node from one increment to the next is a criterion, it might be necessary to release the limit significantly. However, the limit of residual force should not be increased.

3.4 Determination of the Cohesive Parameters

In this section, two ways of determining the cohesive parameters will be presented, namely numerical fitting procedures and direct procedures based on specific tests. Although the fitting procedures are regarded as the standard methods, the direct procedures will be described first since their results may be used as starting values for fitting, in order to reduce the number of fitting runs. Figure 3.3 gives an overview of various independent procedures.

It was already mentioned that two out of the three parameters, T_0, δ_0 and Γ_0, are independent. In this Procedure, the parameters cohesive energy and cohesive

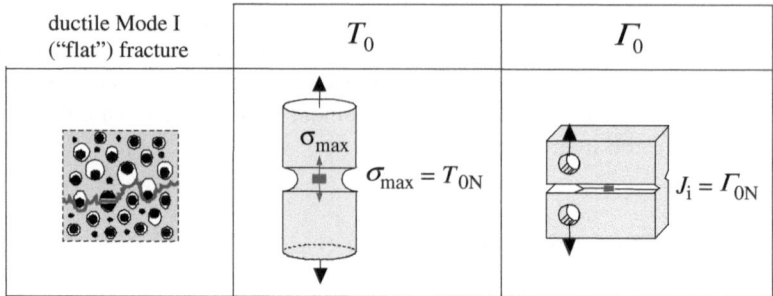

Fig. 3.3 Overview of methods for the determination of the cohesive parameters for ductile metals

strength are suggested to be used, if a direct procedure for their identification is applied. Of course, a fitting algorithm can deal with either set of parameters.

NOTE   It should also be possible to use either of these parameters together with the critical separation, δ_0, which is approximately equal to the crack tip opening displacement at initiation of ductile crack extension. However, practical experience is not yet available.

NOTE   Direct methods for determining cohesive parameters are in principle attractive because they are based on the idea that the thus determined parameters are transferable, i.e. that they exhibit the same values in the specimen used for their determination and in the configuration to be assessed, in particular a configuration containing a crack. In spite of some successful applications, this presumption is not generally valid, wherefore it is recommended to use direct procedures in combination with one of the fitting procedures described below.

NOTE   The cohesive parameters depend on the degree of triaxiality of the stress state (vulgo: constraint) which may vary along the crack front in thick-walled materials and depend on the amount of crack extension. However, in the suggested procedure, they are supposed to be constant, i.e. to have those values that have been determined by the suggested procedure. In Sect. 5.1 hints are given on how these parameters may vary for a given configuration.

3.4.1 Direct Identification Procedures

3.4.1.1 Cohesive Strength

For flat fracture, a tensile test is performed on a round specimen containing a circular notch. A suggested notch geometry is depicted in Fig. 3.4 which also

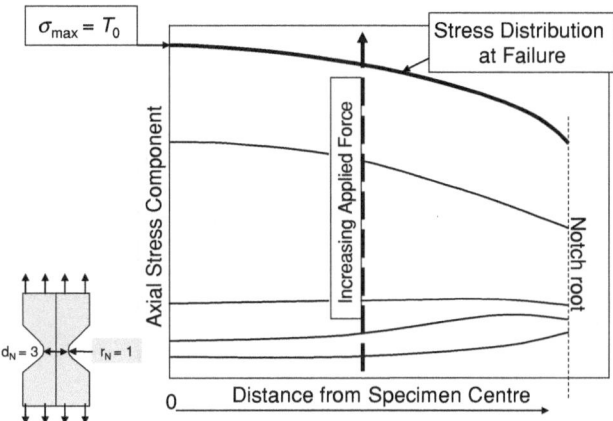

Fig. 3.4 Schematic showing the determination of the cohesive stress on a notched round tensile specimen (d_N and r_N in mm)

shows the procedure for the determination of T_0. From the experiment, the force, F, and the thickness reduction at the notch root, Δb, should be measured. For the incident of fracture, the stress distribution over the specimen's cross section is determined by an elastic–plastic finite element analysis. It is important to note that the instantaneous cross section at fracture is to be used, which means that continuous measurements of the cross section have to be made during the test to capture that condition which corresponds to maximum applied force. The simulated $F(\Delta b)$ curve should meet the experimental one until the incident of fracture, where the experimental curve suddenly drops. At this point the stress distribution over the specimen's cross section is determined from the simulation. The maximum value of that stress distribution is set equal to the cohesive strength, T_0.

For slant fracture, no general direct procedure exists. In the appendix, examples show that the cohesive stress can be determined by means of a tensile test on a flat specimen. In this case, a uniform stress state can be assumed across the specimen's cross section. Therefore, T_0 is given by dividing the force at fracture by the specimen's instantaneous cross section. The cross section of the specimen can be measured post mortem. However, in other tests, the identified value was significantly too small. As a consequence, the directly determined procedure does probably not lead to the correct value, but in any case servers as a starting value for further numerical optimization, Sect. 3.4.2.

3.4.1.2 Cohesive Energy

The cohesive energy for **flat fracture** is set equal to the J-integral at crack initiation, J_i. The determination of this quantity requires careful experimentation.

Fig. 3.5 Determination of J_i
by intersecting the R-curve
with the critical stretch zone
width, SZW_c

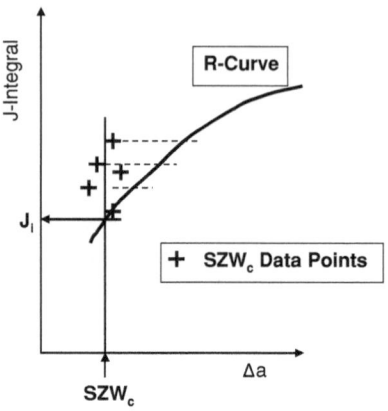

It follows the procedure outlined in the international standard ISO IS 12135 [1] or the GKSS test procedure EFAM GTP 02 [2]. On at least three broken specimens exhibiting ductile tearing the critical stretch zone width, SZW_c, has to be determined in a scanning electron microscope. The intersection of a vertical line representing the average stretch zone width in a $J-\Delta a$ diagram with the initial part of the $J-\Delta a$ curve defines J_i, Fig. 3.5. On the same specimens the amount of crack extension needed for the determination of the $J-\Delta a$ curve can be determined; this is indicated in Fig. 3.5.

A direct technique for the determination of the cohesive energy under **slant fracture conditions** has not yet been developed. However, general rules for determining the crack initiation do apply also for thin-walled structures. Namely, the authors have experience in identifying the crack initiation using the potential drop technique and the multi-specimen method for evaluating J_i, see for example [3], where the crack initiation of an Al5083 H321 was determined as $J_i = 10 \pm 3$ kJ/m^2. Even though this is not sufficiently accurate for a cohesive model analysis, it serves as a reasonable starting value for the numerical optimisation described in Sect. 3.4.2.

3.4.2 Identification Procedure Using Numerical Optimisation

A two parameter optimisation method is suggested in the present procedure. The experiments for the parameter identification should be performed on a standard fracture specimen. The size of the specimen is of minor importance, and thus a small specimen is recommended. The stress state of the specimen for parameter identification should be similar to that of the structure to be assessed in order to assure the same failure mechanism. In addition, if the parameters depend on triaxiality, the values identified are not valid for other stress states. Therefore, in general if the structure under consideration is a thick-walled one (such that the

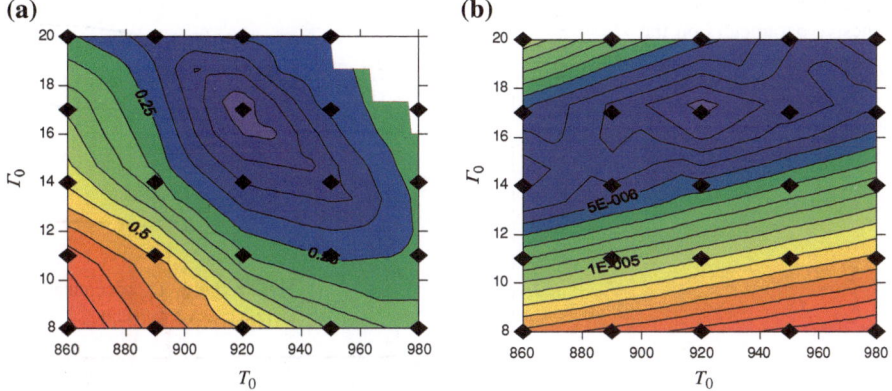

Fig. 3.6 Contour plot of the error between experiment and simulation for an M(T) specimen. **a** error value based on R curve, **b** error value based on load-elongation curve

conditions at the crack tip can be assumed to be plane strain), the specimen for parameter identification must be a plane strain specimen as well. On the other hand, if a thin-walled structure is to be investigated, a sheet panel must be used for parameter identification, since the values for high triaxiality are not applicable.

It is necessary to measure the force, F, a deformation, e.g. CMOD, v_{LL} or even CTOD/δ_5, and the crack extension, Δa. From these values, a force–elongation curve and an R-curve based on J or CTOD can be extracted.

For the numerical optimisation, first an initial set of parameters must be estimated. This can be done by the methods described in Sect. 3.4.1.

It is highly recommended to use an R curve for the parameter identification instead of a force–elongation curve, for the following reasons:

1. It is less sensitive to geometrical differences. For example, the initial crack length needs to be accurately measured, if a force–displacement curve is used, whereas specimen size and initial crack length are not very important for the R-curve.
2. The optimization procedure can find a pronounced local minimum, i.e. the sensitivity of the parameters is high, if the R curve is used. The force–displacement curve leads to a very elongated minimum valley, which is shown in Fig. 3.6.
3. The effect of the parameters can be clearly distinguished in the R curve, which is especially useful for manual parameter identification procedures. Further details on this issue are given in Fig. 3.7 and the respective text.

Starting values for numerical optimization: The starting value for the cohesive energy should be taken from a J R-curve as the value for crack initiation, J_i. If the R-curve is given based on CTOD, then the CTOD value at initiation can be taken as a starting value for δ_i.

Fig. 3.7 Effect of the
cohesive parameters on the
R-curve

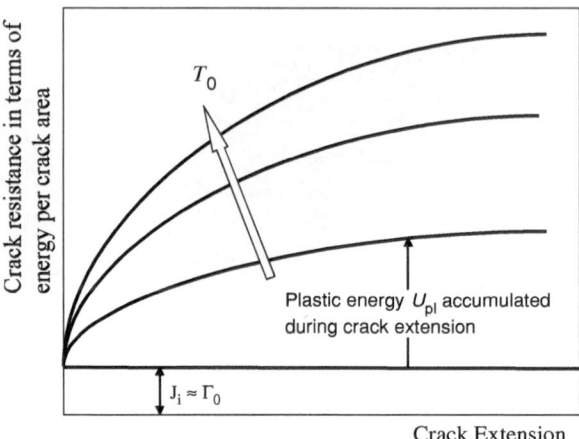

NOTE   If an R-curve is not available at all and the use of a force–displacement
curve cannot be avoided for numerical optimization, the displacement
measured in the experiment should be as close to the crack tip as possible
in order to decrease the amount of continuum deformation on the
displacement. The starting value for the Γ_0 can then be set equal to a J_i
value from literature. If this is not available either, however, a K_{Ic} value
instead, the equation $\Gamma_0 = K_{Ic}^2 / E$ (actually valid for linear elastic
materials only) serves as a (very rough) first estimate

The cohesive strength should initially always be taken from an experiment
according to Sect. 3.4.1. If such an experiment is not available, then a starting value
can be set equal to the stress at failure of the tensile bar, as long as no localised
necking occurs in the specimen during the test. If necking occurs, the cohesive
strength must be higher than the strength of the tensile bar converted in true stress:

$$T_0 > R_m(1 + \varepsilon_m) \tag{3.3}$$

Usually, e.g. for structural steels, it is three times the yield strength, but this is a
very rough estimate, values are also reported between 2.5 and 5 σ_Y.

For the optimisation procedure itself, the following three procedures are
recommended:

3.4.2.1 Trial and Error

Although time consuming, this is probably the most frequently used procedure for
identifying the cohesive parameters. Depending on the user's experience, a few up
to some 20 or more simulations are necessary to achieve a reasonably good
approximation of the experiment. Of course, the user should be able to anticipate
effects of variations in the parameters T_0 and Γ_0 on the numerical solution. It is

highly recommended to use a comparison of the simulated with the experimentally determined R-curve as a criterion for the quality of the numerical solution, since the separate effects of T_0 and Γ_0 on the simulation can only be distinguished in this curve; Fig. 3.7 shows schematically how the cohesive parameters affect an R-curve. Beyond this, an objective target function for the minimisation of the error should be used.

The criterion for the quality of a numerical solution is defined by the user, and different definitions will obviously result in different rankings; therefore it is recommended to apply an objective criterion as described in the following paragraph.

3.4.2.2 Numerical Optimisation by Error Minimisation

Two types of methods have gained general recognition: Gradient methods and evolutionary algorithms. Since the design space is continuous and smooth in a confined range, the chance of getting the global minimum from reasonable starting values make gradient methods the preferred method for the identification of the cohesive parameters.

The objective criterion for minimisation is defined by a target function, which should be the error norm between the experimental and the numerically determined R-curves. Since experimental curves usually consist of a large number of points, the curve can be strongly reduced, e.g. 5–10 points at equidistant crack extension values up to the maximum value of interest. The error is calculated from these values by

$$f = \sqrt{\frac{1}{n-1} \sum_{i=1}^{n} \left(y_i^{\exp} - y_i^{sim}\right)^2} \tag{3.4}$$

with n being the number of sampling points and $y_i^{\exp}, y_i^{sim} y_i$ the respective J or CTOD values for experiment and simulation. It is recommended to use the absolute difference between experiment and simulation and not the relative ones, $\left(y_i^{\exp} - y_i^{sim}\right)/y_i^{\exp}$, since the latter gives more weight to the initial range of crack extension, where the error due to uncertainties in the crack extension measurement is larger.

A common problem for automated optimization arises if the simulation does not converge. The resulting differences cannot be calculated in this case and the error calculation is erroneous. Therefore it is highly recommended to check whether at least the start value is reasonable and the simulation converges in this case. If the FE software allows for addition solution control parameters, these should be used, see Sect. 3.3.2.

As the name says, gradient calculations are necessary for gradient algorithms. However, such gradients of the parameters, df/dT_0 and $df/d\Gamma_0$, are not available if commercial FE solvers are used. The differentiation must be substituted by a calculation of the differential quotient:

$$\frac{df}{dT_0} \approx \frac{f(T_0 + \Delta T_0)}{\Delta T_0} \; ; \quad \frac{df}{d\Gamma_0} \approx \frac{f(\Gamma_0 + \Delta \Gamma_0)}{\Delta \Gamma_0} \tag{3.5}$$

The values ΔT_0, $\Delta \Gamma_0$ must not be too large or too small, as a rule of thumb $\Delta T_0 = [0.5 \ldots 2]$ MPa, $\Delta \Gamma_0 = [0.2 \ldots 1]$ kJ/m^2 should suffice. If the differences are too small, numerical inaccuracies become dominant since the crack extension is not continuous in FE analyses; if they are too large, problems arise when approaching the local minimum. In order to save computation time, a line search algorithm should be used after each optimization iteration, since the line search does not need a gradient calculation.

3.4.2.3 Neural Network

In contrast to the numerical optimisation procedures mentioned in Sect. 3.4.2.2, there is no need to define an error measure. The procedure for the application of neural networks is partitioned into two parts: First the network is trained using the results of various numerical simulations (e.g. a number of points on the J R-curve) within the range of possible cohesive parameter values, and then the experimental J R-curve is fed into the neural network in a second step to obtain values for the cohesive parameters.

a) *Fundamentals*: An artificial neural network (ANN) can be used for solving complex inverse problems in computational mechanics, see e.g. [4, 5]. The underlying theory is fairly simple: An input vector, x_i, is transformed into the output vector, y_i, by an interconnected network of neurons arranged in layers as shown in Fig. 3.8. In a first step, which is called the feed forward step, the processing of the neural network is from left (input) to right (output). Each single neuron has multiple inputs, y_i, drawn as arrows in the figure, and a single output vector, y_j. The activation v_j of the neuron is a linear combination of all inputs multiplied with a specific synaptic weight w_{ij} plus a threshold θ_j:

$$v_j = \theta_j + \sum w_{ij} y_i. \tag{3.6}$$

Then the output y_j of the neuron is computed from a sigmoidal activation function

$$y_j = f(v_j) = \frac{1}{1 + e^{-v_j}} \tag{3.7}$$

which is distributed to all neurons in the following layer.

The ANN is trained by multiple sets of pointwise known correlations between the input and output vectors of interest in order to determine the synaptic weights and thresholds. The internal minimisation strategy behind that identification

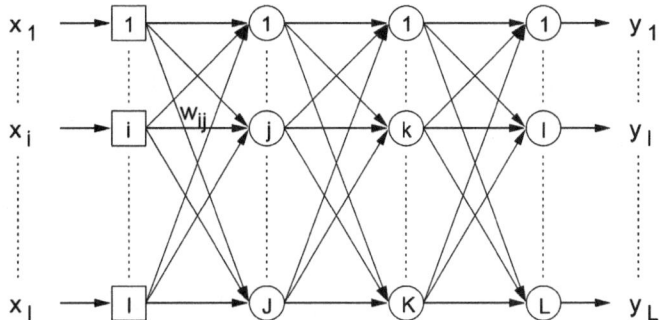

Fig. 3.8 Sketch of a multilayered artificial neural network, [4]

procedure is called "resilient back-propagation", described in [4]. Using the well-trained neural network, approximate solutions can be calculated for given inputs and unknown output vectors.

b) *Setup of the ANN for parameter identification*: The ANN can be used to develop a relation between the shape of a curve of experimentally measurable quantities such as force versus elongation, and the cohesive parameters, T_0 and Γ_0. Therefore, the inputs for the neural network are a small number of sampling points. These might be one of a mixture of the following:

- Values of crack resistance in terms of the *J*-Integral or CTOD at distinct crack extension values.
- Particular quantities from the experiment, e.g. the maximum force, F_{max}, or the *J*-integral or CTOD at crack initiation.
- Several force values F_i at distinct elongations.

The length of the input vector should be kept small, in general 4–6 values should suffice as input.

The two cohesive parameters, either T_0 and Γ_0 or T_0 and δ_0, are assigned to the output neurons (y_1, y_2). The total number of layers is chosen to be three, i.e. only one hidden layer has been inserted. During the training phase one must keep in mind, that the resilient back propagation tries to reproduce the known input–output relations in an optimal way, but this might lead to a so-called overtrained network, i.e. other values cannot be reproduced at all. This behaviour can be avoided if some relations are not used for training but for validation only. With these relations one can see whether the error for other simulations than those used for training is reasonably low as well.

c) *Application of the ANN*: Training sets are generated from a number of finite element simulations with variation of the cohesive properties, providing the respective structural response.

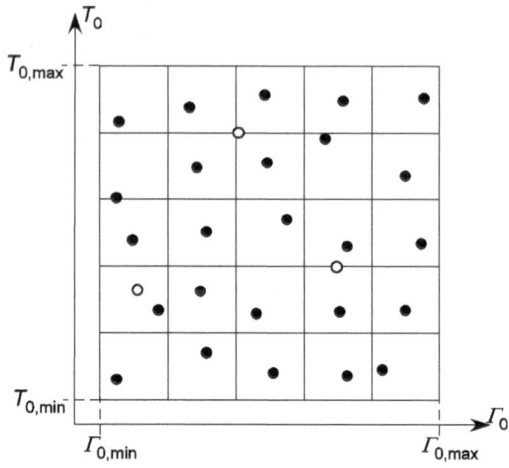

Fig. 3.9 Values taken randomly for the training of an ANN in the range $[T_{0,min}/\Gamma_{0,min};\ T_{0,max}/\Gamma_{0,max}]$, which is divided in 5×5 equal subregions. The hollow symbols denote simulations used for validation

An application to the ANN for identification of cohesive model parameters for both normal and tangential parameters is given in [7]. In general, the following recommendations apply to the choice of the training simulations:

1. The minimum and maximum values of the cohesive parameters must cover the final values, since the ANN cannot extrapolate to values outside the trained regime. Therefore the region must not be too small. The maximum value for the cohesive energy can easily be ten times larger than the minimum value. The maximum value for cohesive strength should be 50 % larger than its minimum value.
2. Due to the nonlinear dependence of the R-curve on cohesive parameters, one should use a sufficient number of simulations in the training. It is recommended to not use less than 25 simulations for training.
3. At least 10 % of the training simulations should be run for validation.
4. The parameters used for the training should not be arranged in a regular pattern, but identified randomly. In order to cover the whole range, one can subdivide the whole training range in smaller regions and take random parameters out of one the subregions shown in Fig. 3.9.

After training, the Artificial Neural Network is used to determine the cohesive parameters from the neural network output for given experimental data at the input. It is important to run a simulation using the parameters retrieved by the ANN such that it is ensured that these values in fact reproduce the experimental results.

References

1. ISO IS 12135 Metallic materials—Unified method of test for the determination of quasistatic fracture toughness. Int. Organ. Stand. Geneva (2007)
2. Schwalbe, K.-H., Heerens, J., Zerbst, U., Pisarski, H., Kocak, M.: EFAM GTP 02—the GKSS test procedure for determining the fracture behaviour of materials. Report No. GKSS 2002/24, GKSS-Forschungszentrum Geesthacht GmbH, ISSN 0344-9629 (2002)

3. Scheider, I., Schödel, M., Brocks, W., Schönfeld, W.: Crack propagation analysis with CTOA and cohesive model: comparison and experimental validation. Eng. Fract. Mech. **73**, 252–263 (2006)
4. Huber, N., Tsakmakis, C.: A neural network tool for identifying the material parameters of a finite deformation viscoplasticity model with static recovery. Comput. Methods Appl. Mech. Engrg. **191**, 353–384 (2001)
5. Brocks, W., Steglich, D.: Hybrid methods. In: Milne, I., Ritchie, R.O., Karihaloo, B. (eds.) Comprehensive structural integrity, pp. 107–136. Elsevier, Amsterdam (chapter 10.05) (2007)
6. Scheider, I., Brocks, W.: Simulation of cup-cone fracture using the cohesive model. Eng. 29 Fract. Mech. **70**, 1943–1962 (2003)
7. Scheider, I., Uz, V., Huber, N.: Applicability of a cohesive model to fracture of thin-walled structures: parameter identification and thickness dependence. Eng. Fract. Mech. (2012, in preparation)

Chapter 4
Applications

Apart from the application to pre-cracked bulk materials under static loading, the cohesive model can be applied to almost every problem of the integrity of materials and structural components. The model is gaining increasing interest for application, it is in particular ideally suited for large amounts of crack extension and the behaviour of interfaces, such as phase boundaries, coatings, bonded joints, delamination in layered materials, and fibres in matrices as well as the prediction of fracture paths. Crack extension in bulk materials will be shown in some detail whereas other areas of application outside the experience gained at GKSS will only be briefly touched upon in this chapter. They demonstrate the enormous range of problems which can be treated using the cohesive model.

NOTE   It is crucial that the TSL shape used for parameter identification is the same as that used application to components, since a different shape would lead to different parameters

4.1 Damage Free Material

It is worth noting that cohesive models are able to handle damage anywhere in a material, the pre-existence of a crack for modelling damage is not needed. This is of particular interest for concrete and rocks—i.e. very brittle materials—which frequently do not contain initial macroscopic cracks. Modelling of forming processes using the cohesive model is another area of interest; in this case, deformation limits without introducing damage into the component can be predicted. However, it must be kept in mind that the stress state in an uncracked structure is in general very different from the state ahead of the crack tip, i.e. the triaxiality is much lower. Thus the cohesive parameters cannot be determined by direct procedures outlined in Sect. 3.4.1, but must be identified numerically from fitting to

K.-H. Schwalbe et al., *Guidelines for Applying Cohesive Models to the Damage Behaviour of Engineering Materials and Structures*, SpringerBriefs in Applied Sciences and Technology, DOI: 10.1007/978-3-642-29494-5_4, © The Author(s) 2013

experiments, which show a similar stress state as the one investigated. Sufficient experience on this issue is not yet available. A recent work on this topic has been published by Banerjee et al. [1].

4.2 Treatment of Thin-Walled Structures

Mode I fracture with a fracture surface normal to the global loading direction, frequently called *flat fracture*, is usually modelled by 2D plane strain or 3D models.

In thin-walled structures made of ductile metals, upon loading the front of a crack with its plane perpendicular to the global loading direction, starts turning into a plane inclined to 45° to the loading direction, which is usually called *slant fracture*, Fig. 4.1. As already mentioned in Sect. 3.4.1, the cohesive strength can be identified by direct procedures, but no generally applicable procedure is available for the cohesive energy.

Since the crack starts perpendicular to the global loading direction, the correct way would be to divide the ligament containing cohesive elements into one region, where the cohesive parameters are determined for flat fracture, and a second one with the cohesive parameters for slant fracture. However, since the first region is rather short, and its length can be approximated by the thickness of the sheet, the first region might therefore be neglected for larger structures. This simplification has been used in almost all publications, see e.g. [3], where rather small specimens, namely C(T) specimens with W = 50 mm were evaluated. However, if even smaller specimens are to be simulated, the contribution of the first region to the behaviour of the specimen might not be neglected any more, as has recently been investigated by simulations of Kahn specimens [4]. There the ligament was divided into one region, where the cohesive parameters are determined for flat fracture, and a second one with the cohesive parameters for slant fracture. The values of the cohesive parameters for these two regions are generally different, and not much experience has been gained about a specific relation between the parameters in the two regions. As a rule of thumb it can be stated that both the cohesive strength and the cohesive energy for the slanted part are lower than the ones for flat fracture, see e.g. [5–7]. In [5] the crack extension of thin-walled fracture specimens failing in a slanted manner was controlled by adjusting the cohesive energy in each element by experimental v_{LL}–Δa curves. By reproducing the experimental data it turned out that the crack initiates with a Γ_0 value, which is equal to the crack initiation J for normal fracture, and then reduces to significantly lower values during the transition to the slanted fracture mode. After reaching the fully slanted region, the values remain almost constant again, see Fig. 4.2.

Commercial finite element codes in general do not allow defining a variable cohesive energy controlled by a given crack resistance curve. However, in order to perform reliable crack extension simulations, the following methods can be applied for modelling thin-walled structures:

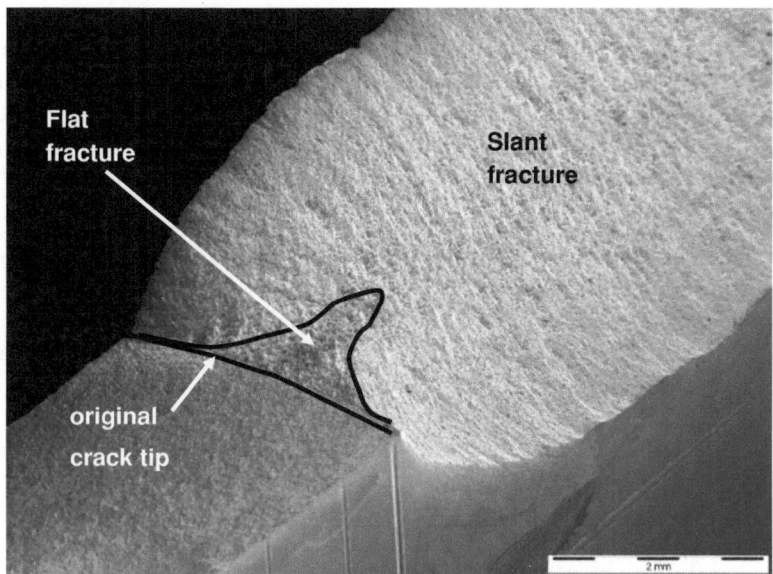

Fig. 4.1 Transition from normal to slant fracture in thin walls; from [2]

Fig. 4.2 Development of the cohesive energy in the transition region from flat to slant fracture, identified by controlling the crack extension in the simulation with an experimental v_{LL}–Δa curve, from [5]

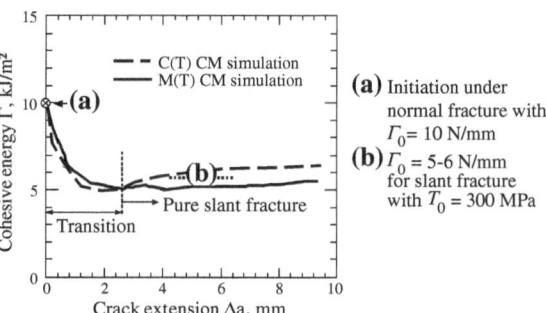

(a) Initiation under normal fracture with Γ_0= 10 N/mm

(b) Γ_0 = 5-6 N/mm for slant fracture with T_0 = 300 MPa

- If a thin-walled structure is modelled by shell elements, which is the most common technique, the cohesive model can only show normal separation, since any inclination in thickness direction does not exist in the finite element model. Since any three-dimensional stress state ahead of the crack tip is not well represented, the cohesive parameters should be called *effective cohesive parameters*.

- If 3D models are used, one can mesh the actual inclined fracture surface, but this is rather difficult, since a transition region, where the crack deviates from its original direction is necessary in this case. In addition, complex mixed mode conditions are involved: The slanted fracture surface contains Mode I/III conditions, and the transition zone has a very complex Mode I/II/III mix.

- Alternatively, the slant fracture can be modelled by projecting its actual sur-face onto a plane perpendicular to the wall, thus mimicking flat fracture. Of course, from a mechanics point of view, due to this approximation of the real process, the stresses acting on the cohesive elements are no longer the real ones. This way a flat fracture is modelled for Mode I conditions. The cohesive parameters determined by such procedure should be called *effective cohesive parameters*.

The first option is the one, which is recommended for general use. For this option the proposed method for identifying the cohesive strength, see Clause 3.4.1, is applicable. The cohesive energy should be fitted to experimental results using one of the methods described in Clause 3.4.2. More details about additional issues on thin-walled structures are given in [8].

NOTE   The user should keep in mind that even though cohesive elements are applicable to 2D structures by most of the commercial codes, sometimes they do not take the thickness reduction in structures under plane stress conditions into account and thus cannot be used for crack extension analyses with plane stress elements

NOTE   Investigations [9] have shown that a thickness dependence of the simulated R curve is shown even in structures where this cannot be observed in experiments, if the simulation has been performed using the third option. Therefore, the parameters identified for a sheet material using the third option cannot be transferred to the same material with different thickness

4.3 Simulation of Crack Extension (*R*-Curves)

The simulation of crack extension is one of the most frequent applications of the cohesive model. Simulation of the *R*-curve using the cohesive model allows the determination of instability conditions of a structural component and of the correlation between applied loads and a suitably selected deformation.

During the migration of the crack through the material, the conditions of decohesion within the process zone local to the crack tip remain constant and are represented by the cohesive parameters. This is in line with the observation of the crack tip opening angle, CTOA (except for an initial transient range as shown in Fig. 4.2). Measuring this angle shows a constant value during crack extension, apart from an initial transitional behaviour right after initiation of crack exten-sion. Both, the transition region as well as the steady state regime can be reproduced with the cohesive model using constant parameters [3]. In contrast to this physically plausible fact, the resistance as measured using standard test procedures increases with the amount of crack extension. The reason is that the actual standard test methods use global parameters such as the tress intensity factor, the *J*-integral, or the crack opening displacement at the position of the pre-crack tip to measure the resistance to crack extension, which are highly dominated by plastic deformation which in turn has nothing to do with the actual decohesion process, whereas the cohesive energy represents the energy consumption and hence the material's resistance to crack extension in the process zone local to the crack tip, which is constant during the motion of the crack through the material. The fact that in contrast to the traditional *R*-curve parameters, the cohesive parameters can be assumed to be constant during crack extension makes them highly attractive for simulating crack extension in a structural component.

NOTE   Although slight changes in the stress field during crack extension are observed which may have an effect on the cohesive parameters, constancy of these parameters can be assumed for practical purposes

An actual simulation of crack extension in a structural component follows the procedures outlined in Chap.3, for details see Appendix 1.

4.4 Interfaces

Interfaces are the natural playground for the cohesive model since the model is by definition based on properties of an interface. In practice, a number of material configurations can be reduced to material phases separated by an interface, suchas

- Joints (weldments, adhesive bonds etc.).
- Coatings.
- Reinforcements within a material (fibres, particles, etc.).
- Laminates.
- Adjacent phases of metallic materials.

to name a few. In reality, the interface is not a mathematical fiction, but a layer with a finite, albeit very small, thickness the properties of which are represented by the cohesive model.

When treating cracks in interfaces it may be expected that the crack runs along the interface to final failure. This, however, is not always the case. Other failure mechanisms usually compete, so the crack may deviate into one of the phases adjacent to the interface. This issue is considered in Sect. 4.4.3.

4.4.1 Welded Joints

A welded joint represents a complex detail of a structural component in that the material properties may vary substantially across the weld, details depending on the materials welded together and the welding process. Metals can be joined by a

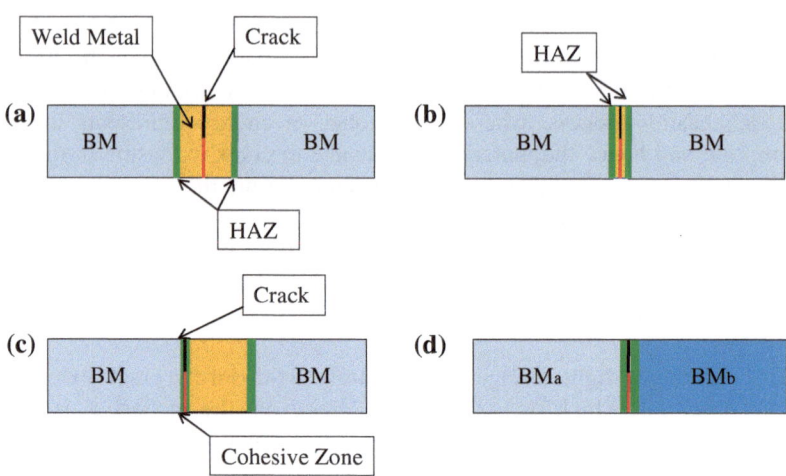

Fig. 4.3 Modelling of welded joints: Approximation of the material regions (BM = base metal, HAZ = heat affected zone); **a** Typical fusion weld with crack in weld metal; **b** Narrow gap weld with crack in weld metal; **c** Same weld as in Fig. 4.3 a with crack in HAZ; **d** Two base metals joined by friction weld

wide selection of welding procedures. Two metallic pieces welded together using a fusion welding procedure is probably the most commonly used welded joint. This kind of joint consists of two base plates made of the same material (or two different base metals if two different materials are joined); the weld metal, formed by the molten and then solidified consumable, and the heat affected zone in each of the two base plates. The thermal cycles of the welding process create a complex microstructure with widely varying mechanical properties; for example, the heat affected zone (HAZ), representing that volume of the base plate which is adjacent to the weld metal undergoes a suite of thermal cycles, ending up with a range of microstructures and hence mechanical properties, so that even the term "heat affected zone" is frequently used in plural. This situation makes any attempt to model a weld a difficult task, thus requiring simplifications to make the problem tractable, see e.g. [10–12].

Figure 4.3 depicts a variety of typical welded joints, and in the context of interfaces the question arises as to where in these welds interfaces can be identified. Fusion welds exhibit a fusion line (in fact it is a fusion surface, in micrographs only the trace of the fusion surface is visible) which separates the weld metal from the heat affected zone.

Because of its importance, in Fig. 4.3a the above mentioned kind of weld is depicted, with a crack within the weld metal, although the crack is not immediately related to the interface. However, the interface has an indirect effect on the behaviour of the crack since the yield strength mis-match between the weld zones usually present in a weld affects the stress and strain fields around the crack [13]. If the weld is sufficiently narrow like in a narrow gap weld, Fig. 4.3b, then the proximity of the crack to the fusion zone and the yield strength mis-match may give rise to an interaction between crack driving force and crack resistance which attracts the crack from the weld metal to either the heat affected zone or the base metal. This has to be accounted for when the cohesive elements are placed along the expected crack path(s). A similar situation is given in Fig. 4.3c where the crack lies in the heat affected zone, it has the choice to deviate to either side, see Fig. 4.7 which shows a crack deviating from the fusion line into the yield strength under matched weld metal. Figure 4.3d shows an even more complex arrangement given by a friction welded joint of two different base materials where the crack has in principle five different paths: along the fusion line, into the heat affected zones of either base material, or into either base metal. In practice, however, experimental experience with actual joints may limit the number of choices.

Modelling of a structure with a crack in the weldment according to Fig. 4.3 may be a challenging task due to the yield strength mismatch and the resulting crack path deviation. For a reliable but affordable case study, several different cases must be considered.

(A) The crack initiation and initial direction are important. That is, residual strength of the structure is mainly determined by the crack path and the crack does not extend much before instability. In this case, it is important to model the material inhomogeneity with high accuracy. Crack path issues must be taken into account, see Sect. 4.5.

(B) Long crack extension is achieved before final failure. In this case the initial crack position is less important than the location of the main part of the crack extension. If for example the crack starts from the centre of the weld but turns into the heat affected zone after crack initiation and has its main propagation there, the initial crack tip can be located in the HAZ. It is important that the configuration used for the identification of the cohesive parameters is similar to that of the structure under investigation—e.g. material thickness, tension or bending loading, biaxial loading, size of the component.

(C) If both the crack extension and structure geometry are large with respect to the weld width, which is for example the case for laser welded thin sheets, the whole weldment can be replaced by a single cohesive line. In this case the configuration used for the identification must meet that of the structure even more closely, that is, all weld parameters, sheet thickness, etc. must be the same in all simulations. Inaccuracies can be accepted, if in return it is possible to model very large structures such as car wings, hoods and airplane fuselages.

4.4.2 Delamination

Structural components in service are always subject to damage. Therefore, non destructive inspections have to be carried out on critical components, i.e. those components whose failure may lead to severe damage to the complete system. For components made of metallic materials, highly sophisticated techniques have been developed and in combination with predictive tools, inspection intervals can be set up and successfully used for safe operation. However, when applying fibre reinforced composites to structural components, then the situation is much less advanced. In these materials, delamination is the main damage mechanism which is difficult to detect, and models for routine simulation are still lacking. An example for addressing this problem is provided by [14] who developed a finite element strategy with relatively large elements in order to treat large structures with a limited number of elements.

4.4.3 Coatings

Structural materials are increasingly given coatings, to provide either protection against environmental attack and increased wear properties, or adding functions such as optical properties and health monitoring.

Coatings may consist of a single layer, or of several individual layers, making the computational model complicated. Multilayer coatings may, therefore, be simplified such that single coatings result, see e.g. [15]. These authors did a parameter study on the behaviour of thermal barrier coatings and developed criteria for crack extension along the interface or coating fracture.

The modelling of coatings may include two aspects: The delamination of the coating from the substrate, see Sect. 4.4.2, but also breaking of the coating, which happens most likely after some amount of delamination. For the latter case it is necessary to insert cohesive elements vertically to the surface of the coated structure. Since it is not known in advance where breaking of the coating occurs, this is a crack path prediction problem, see Sect. 4.5.

Coatings are often very thin compared to all other geometrical dimensions, usually much smaller than a millimetre. This induces that the prerequisites of continuum mechanics are not met anymore. In addition, the roughness of the interface between substrate and coating may be too large to be neglected in the simulation. That is, the user must ensure that a single straight interface is a valid approximation.

4.5 Prediction of Crack Path

The main drawback of the cohesive model and its implementation using interface elements is that the crack path must be prescribed by the user. However, several research groups have tried to predict the crack path using cohesive modelling as well, e.g. [16]. If a few alternatives exist for a crack to extend, then it is easy to define these possible crack paths as cohesive zones. This is usually the case in the neighbourhood of mechanical interfaces, which the crack may either cross or run along. The two major types of interfaces are geometrical interfaces, e.g. a stiffener

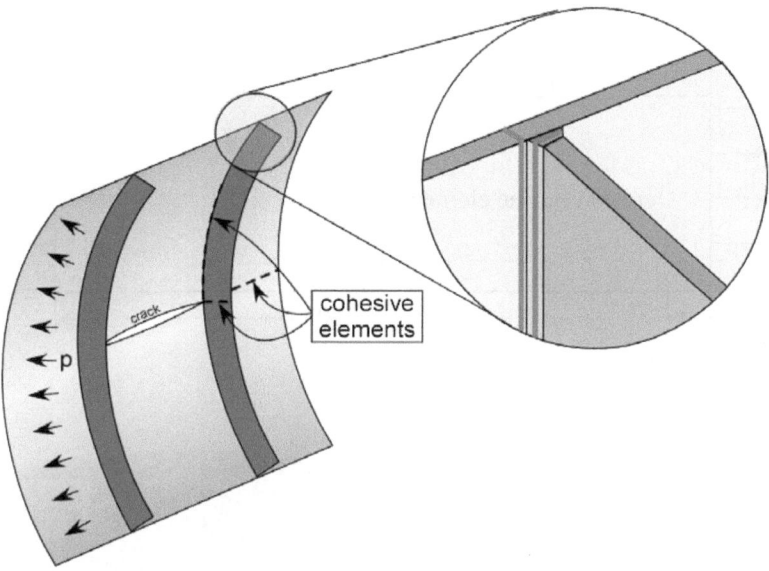

Fig. 4.4 Crack path prediction (either through or along the stiffener), from [17]

attached to a panel, and material interfaces, across which the properties change. An important example for such cases is given by welded joints, see Sect. 4.4.1. If a crack approaches such a heterogeneity, crack path deviation is likely to occur.

An example for crack deflection at a geometrical discontinuity, namely a cylindrical shell with a circumferential stiffener and a longitudinal crack approaching the stiffener, taken from [17], is shown in Fig. 4.4.

A crack approaching an interface can either penetrate it or run along that interface. This configuration was investigated in [18] concerning elastic mismatch, see Fig. 4.5. On the microscale, the interface problem was investigated e.g. in [19], when a crack propagates through the metal of a metal matrix composite and approaches the (elastic) fibre. In this case, the fibre may debond or break, which is a bifurcation problem as well, see Fig. 4.6.

The effect of crack deflection to an arbitrary angle and thus the prediction of a completely unknown crack path has also been investigated by several researchers for mismatched solids. Among others, simulations using finite element meshes that consist of a pattern of four triangles forming a quadrilateral with cohesive elements between all element edges were performed in [20, 21], and [22]. Such an element pattern has already been shown in Fig. 2.7. Since the number of elements and nodes increases significantly by this meshing technique, the region of elements should be chosen as small as possible. For example, in [20] a crack propagating along an interface between a base material and a softer weld seam is modelled. In this case it is assumed that the crack deviates into the weld metal and thus only that region has been modelled by the triangular pattern, see Fig. 4.7.

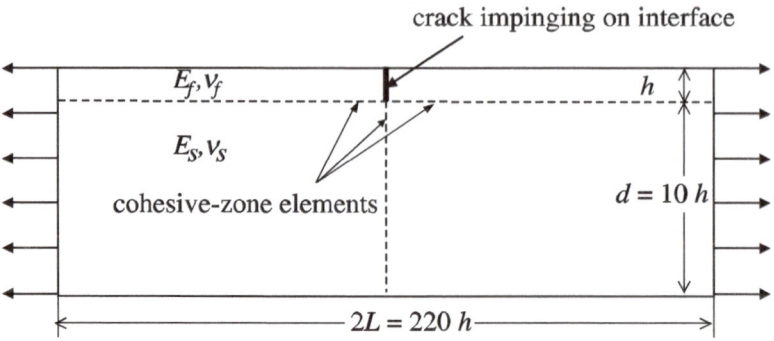

Fig. 4.5 Material interface concerning elastic mismatch, from [18]

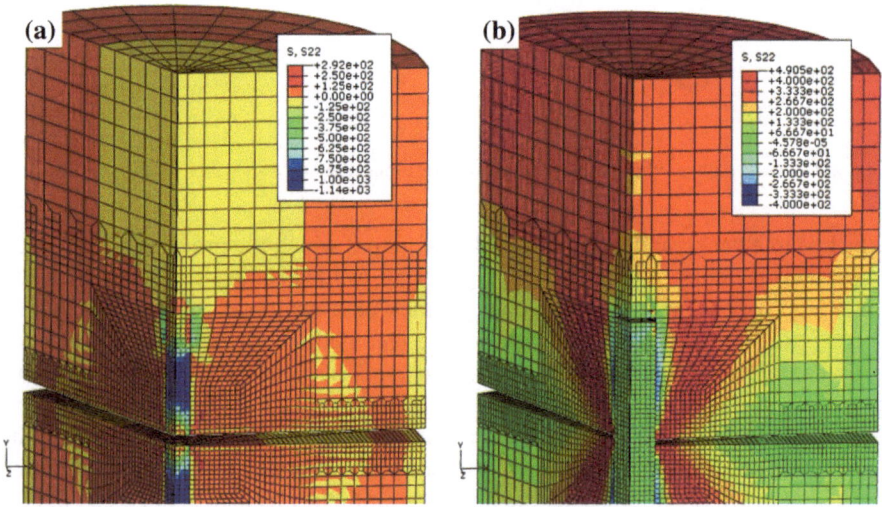

Fig. 4.6 Material interface at the microscale: Fibre reinforced composite containing a crack, which approaches the fibre and may either break (**a**) or debond (**b**) it

4.6 Time Dependent Effects

The mechanical behaviour of materials may depend on the time explicitly (creep effects) or on the rate of the applied deformation, $\dot{\delta}$. In addition, environmental effects like corrosion, diffusion, etc., depend on time by nature. All of these problems have been investigated by various researchers. In general the procedure is the same: The TSL in these cases does not only depend on the separation itself, but on other quantities as well: $T = f\left(t, \delta, \dot{\delta}, \varsigma_i(\mathbf{x}) \dots\right)$, where ς_i are field quantities depending on the position, $\mathbf{x}$. For example, this could be the hydrogen concentration in the case of stress corrosion cracking, see Sect. 4.6.3. An easy and

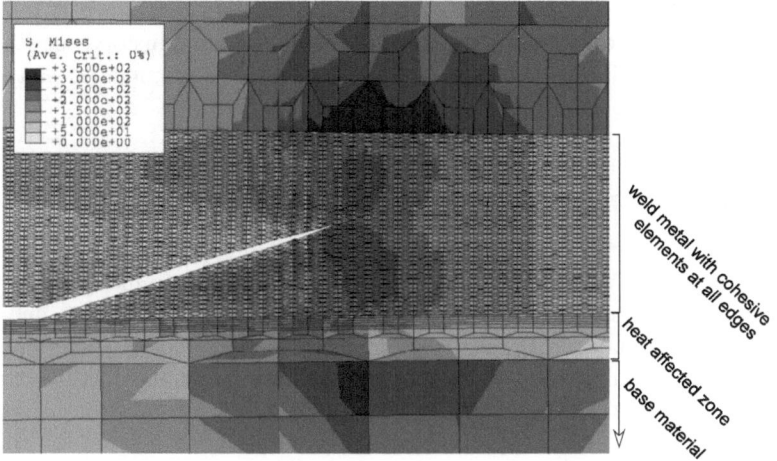

Fig. 4.7 Modelling of crack path deviation at a material interface (welded joint) allowing the crack to propagate away from the bond line; from [20]

versatile method of including this dependency is to make the cohesive parameters, T_0 or Γ_0, dependent on these quantities.

4.6.1 Rate Dependent Formulations

Three types of rate dependent formulations are often distinguished:

- Explicit rate dependency.
- Viscoplastic behaviour.
- Viscoelastic behaviour.

The explicit rate dependency is the simplest approach, in which the TSL is usually written with two terms, one of which depends on the separation itself, and the other on the separation rate:

$$T = f_1(\delta) + f_2\left(\dot{\delta}\right)$$

or   (4.1)

$$T = g_1(\delta) g_2\left(\dot{\delta}\right)$$

A cohesive law with a very simple form of this type, namely

$$T = T_0 + \eta\dot{\delta} \quad \text{for} \quad \delta < \delta_0$$ (4.2)

has been proposed by [23].

Viscoelastic or viscoplastic laws are more complicated since viscoelastic extensions usually contain a time integral of the deformation history, see e.g. [24], whereas the separation must be split into an elastic and a viscoplastic part for the viscoplastic formulations, [25].

4.6.2 Dynamic Fracture

Dynamic problems were investigated by several groups. Even though it is a rate-dependent formulation, see above, it is treated in a separate clause due to its importance for metallic materials (e.g. crash tests, etc.) and its easy implementation. Besides defining a dependence on the separation rate, the cohesive strength and energy may also depend on the strain rate, which is available in the adjacent continuum element. This approach was applied in [26], where the cohesive properties were identified at high speed tests as

$$\left.\begin{array}{l} T_0 \approx C_1\sigma_0\left(\frac{\dot{\varepsilon}}{\dot{\varepsilon}_0}\right)^m \\ \Gamma_0 \approx C_2\sigma_0\left(\frac{\dot{\varepsilon}}{\dot{\varepsilon}_0}\right)^m \end{array}\right\} \quad \text{for} \quad \dot{\varepsilon} > \dot{\varepsilon}_0 \tag{4.3}$$

with additional parameters C_1, C_2, $\dot{\varepsilon}_0$, and m.

NOTE   If commercial codes are used, the user must be aware that dependencies on additional field quantities are usually not possible, at least not for quantities, which do not exist in the cohesive element itself but only in the adjacent continuum element (like strains or strain rates).

4.6.3 Stress Corrosion Cracking

The material degradation due to environmental conditions and additional mechanical stresses is typical of the hydrogen diffusion in stress corrosion cracking. The driving force is the hydrogen concentration, and therefore the cohesive properties depend on this quantity. A few references exist from different groups, e.g. [27, 28] etc., where the cohesive strength is reduced by the concentration, whereas the critical separation is not affected. The simplest form of cohesive strength reduction is given by

$$T_{0,\text{SCC}} = T_0(1 - \alpha C/C_{\text{env}}) \tag{4.4}$$

i.e. a linearly decreasing function with additional parameters α and C_{env}, see e.g. [29]. The note given in Sect. 4.6.2 concerning the dependence on additional field quantities in combination with commercial codes holds also for the problem of hydrogen dependence.

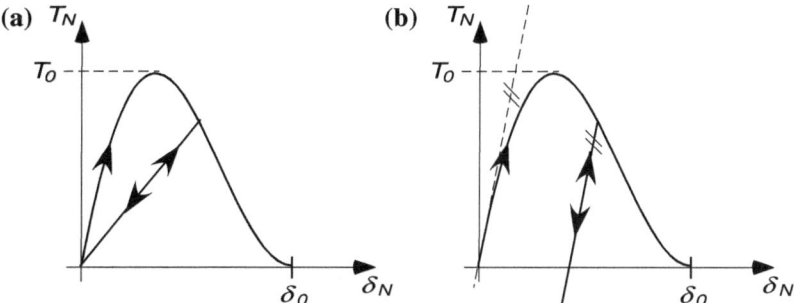

Fig. 4.8 Unloading algorithms for cohesive elements; **a** The separation vanishes when unloaded to the stress-free state; **b** Unloading algorithm leading to some remaining separation

4.7 Unloading and Reversed Loading, Fatigue

If a load path other than a monotonous one is investigated, two additional characteristics of the cohesive behaviour must be taken into account:

Unloading algorithm: This must be implemented in the traction–separation law, i.e. the response to reduction of interface separation or traction acting on the interface.

Damage accumulation: This characteristic effect is only important for repeated loading.

Unloading algorithm

This is important for any unloading or reversed loading. Different characteristics have been described by various authors. The two main differences can be distinguished by the behaviour under total unloading leading back to the stress-free state. Some models then reduce the separation to zero, see Fig. 4.8a, whereas others lead to some remaining separation even in the stress-free state, Fig. 4.8b. The former are more representative of brittle fracture, where microcracks close completely when the load vanishes. Research groups using this type of unloading behaviour are [30] or [31]. A remaining separation represents ductile fracture, where pores and voids, which may grow under loading, cannot close completely, if the stress is released. Such cohesive models are employed e.g. by [29, 32].

Damage accumulation

Saturation occurs easily during cyclic loading, if the load history contains identical cycles, due to the unloading/reloading path, which in general is also equal. If the load does not increase from one cycle to the other, the loading path of the cohesive element will only follow this unloading/reloading path and material degradation does not increase anymore. Two ways to overcome this non-physical behaviour can be found in the literature:

Fig. 4.9 Cyclic loading of
cohesive elements following
a separation law with
different paths for loading
(*solid line*) and unloading
(*dashed line*)

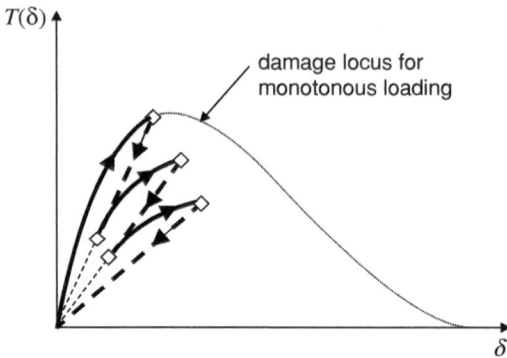

One possibility is to define different paths for unloading and reloading, see Fig. 4.9. This has been described e.g. by [33–35]. In such models the unloading may follow a linear curve, whereas the reloading is based on a quadratic equation.

An additional damage variable, which accumulates over load cycles and changes the behaviour of the cohesive element, has been introduced e.g. by [29] or [36]. In this case, usually the cohesive strength is reduced by the amount of damage.

In both cases the cohesive element may fail even without reaching the initial cohesive strength. However, since the total load cycle history must be followed in the simulation, such analyses are very cumbersome, and therefore they are usually only used for low cycle fatigue up to some hundred cycles.

References

1. Banerjee, A., Manivasagam, R.: Triaxiality dependent cohesive zone model. Eng. Fract. Mech. **76**, 1761–1770 (2009)
2. Scheider, I., Brocks, W.: Simulation of crack propagation and failure in thin walled structures using the cohesive model. In: Loughlan, J. (ed.) Proceedings of the 4th International Conference on Thin-Walled Structures (2004)
3. Scheider, I., Schödel, M., Brocks, W., Schönfeld, W.: Crack propagation analysis with CTOA and cohesive model: comparison and experimental validation. Eng. Fract. Mech. **73**, 252–263 (2006)
4. Cornec, A., Schönfeld, W., Schwalbe, K.-H., Scheider, I.: Application of the cohesive model for predicting the residual strength of a large scale fuselage structure with a two-bay crack. Eng. Fail. Anal. **16**, 2541–2558 (2008)
5. Yuan, H., Lin, G., Cornec, A.: Verification of a cohesive zone model for ductile fracture. J. Eng. Mater. Technol. **118**, 192–200 (1996)
6. Cornec, A., Lin, G.: Kohäsivmodell als lokales Bruchkriterium für duktile Bruchvorgänge. Technical Report GKSS/WMA/98/5, GKSS (1998)
7. Siegmund, T., Brocks, W., Heerens, J., Tempus, G., Zink, W.: Modeling of crack growth in thin sheet aluminium. In: International Exhibition and Conference: Recent Advances in Solids and Structures (1999)

8. Scheider, I., Brocks, W.: Cohesive elements for thin-walled structures. Comp. Mater. Sci. **37**, 101–109 (2003)
9. Scheider, I., Uz, V., Huber, N.: Applicability of the cohesive model to fracture of lightweight structures: parameter identification and thickness dependence. In: 18th European Conference on Fracture, Dresden (2010)
10. Schwalbe, K.-H.: Effect of weld metal mis-match on toughness requirements: some simple analytical considerations using the engineering treatment model (ETM). Eng. Fract. Mech. **67**(257), 277 (1992)
11. Schwalbe, K.-H.: Welded joints with non-matching weld metal—crack driving force considerations on the basis of the engineering treatment model (ETM). Eng. Fract. Mech. **63**, 1–24 (1993)
12. Schwalbe, K.-H., Kocak, M. (eds.): Second International Symposium on Mis-Matching of Interfaces and Welds. GKSS Research Centre, Geesthacht (1977)
13. Schwalbe, K-H, Kim, Y-J, Hao, S, Cornec, A, Kocak, M.: EFAM ETM-MM 96—the ETM Method for assessing the significance of crack-like defects in joints with mechanical heterogeneity (strength mismatch). Report GKSS 97/E/9, GKSS Research Centre, Geesthacht (1997)
14. Turon, A., Dàvila, C.G., Camanho, P.P., Costa, J.: An engineering solution for mesh size effects in the simulation of delamination using cohesive zone models. Eng. Fract. Mech. **74**, 1665–1682 (2007)
15. Yuan, H., Chen, J.: Computational analysis of thin coating layer using a cohesive model and gradient plasticity. Eng. Fract. Mech. **70**, 1929–1942 (2003)
16. Xu, X.-P., Needleman, A.: Numerical simulations of dynamic interfacial crack growth allowing for crack growth away from the bond line. Int. J. Fract. **74**, 253–274 (1995)
17. Scheider, I., Brocks, W.: Residual strength and crack path predictions by the cohesive model. In: Proceedings of the International Conference on Crack Paths, Parma (2006)
18. Parmigiani, J.P., Thouless, M.D.: The roles of toughness and cohesive strength on crack deflection at interfaces. J. Mech. Phys. Solids **54**, 266–287 (2006)
19. Brocks, W., Scheider, I.: Prediction of crack path bifurcation under quasi-static loading by the cohesive model. Struct. Durab. Health Monit. **70**, 1–11 (2008)
20. Nègre, P., Steglich, D., Brocks, W.: Crack extension at an interface: prediction of fracture toughness and simulation of crack path deviation. Int. J. Fract. **134**, 209–229 (2005)
21. Siegmund, T., Fleck, N.A., Needleman, A.: Dynamic crack growth across an interface. Int. J. Fract. **85**, 381–402 (1997)
22. Arata, J.J.M., Kumar, K.S., Curtin, W.A., Needleman, A.: Crack growth across colony boundaries in binary lamellar TiAl. Mater. Sci. Eng. **A329–A331**, 532–537 (2002)
23. Costanzo, F., Walton, J.R.: A study of dynamic crack growth in elastic materials using a cohesive zone model. Int. J. Eng. Sci. **35**, 1085–1114 (1997)
24. Knauss, W.G.: Time dependent fracture and cohesive zones. J. Eng. Mater. Technol. **115**, 262–267 (1993)
25. Estevez, R., Tijssens, M.G.A., Van der Giessen, E.: Modelling of the competition between shear yielding and crazing in glassy polymers. J. Mech. Phys. Solids **48**, 2585–2617 (2000)
26. Anvari, M., Scheider, I., Thaulow, C.: Simulation of dynamic ductile crack growth using strain-rate and triaxiality-dependent cohesive elements. Eng. Fract. Mech. **73**, 2210–2228 (2006)
27. Serebrinsky, S., Carter, E.A., Ortiz, M.: A quantum-mechanically informed continuum model of hydrogen embrittlement. J. Mech. Phys. Solids **52**, 2403–2430 (2004)
28. Falkenberg, R., Brocks, W., Dietzel, W., Scheider, I.: Modelling the effect of hydrogen on ductile tearing resistance of steels. Int. J. Mat. Res. **101**, 989–996 (2010)
29. Scheider, I., Dietzel, W., Pfuff, M.: Simulation of hydrogen assisted stress corrosion cracking using the cohesive model. Eng. Fract. Mech. **75**, 4283–4291 (2008)
30. Camacho, G.T., Ortiz, M.: Computational modelling of impact damage in brittle materials. Int. J. Solids Struct. **33**, 2899–2938 (1996)

31. Chaboche, J., Girard, R., Levasseur, P.: On the interface debonding models. Int. J. Damage Mech. **6**, 220–257 (1997)
32. Scheider, I., Brocks, W.: Simulation of cup–cone fracture using the cohesive model. Eng. Fract. Mech. **70**, 1943–1962 (2003)
33. Roe, K.-L., Siegmund, T.: An irreversible cohesive zone model for interface fatigue crack growth simulation. Eng. Fract. Mech. **70**, 209–232 (2003)
34. deAndrés, A., Pérez, J.L., Ortiz, M.: Elastoplastic finite element analysis of three-dimensional fatigue crack growth in aluminium shafts subjected to axial loading. Int. J. Solids Struct. **36**, 2231–2258 (1999)
35. Yang, B., Mall, S., Ravi-Chandar, K.: A cohesive zone model for fatigue crack growth in quasibrittle materials. Int. J. Solids Struct. **38**, 3927–3944 (2001)
36. Serebrinsky, S., Ortiz, M.: A hysteretic cohesive law model of fatigue-crack nucleation. Scripta Mater. **53**, 1193–1196 (2005)

Chapter 5
Open Issues

As already mentioned in several places, the practical application of the cohesive model is still under development. Therefore, there are items that require further research for the establishment of firm rules in strict procedural form, or to confirm what has been formulated in Chap. 3. In the following sections, several open issues will be discussed in detail, however, it can also be shown that in spite of some quite complex relationships underlying the methods outlined in Chap. 3, these methods represent pragmatic simplifications which can be successfully applied in practice, see the following sections and the worked examples in Appendix 1.

5.1 Determination of the Cohesive Energy by a Direct Procedure

In the present procedure, the direct determination of the cohesive energy is performed by setting it equal to the J-integral at initiation of stable crack extension in order to obtain an independent method for Γ_0, without the need for numerical curve fitting. However, only under extreme small scale yielding conditions can J_i be equal to Γ_0. Otherwise, the plastic zone becomes so large that the energy consumed at crack initiation becomes much greater than the energy needed for decohesion. This is illustrated in Fig. 3.6 where the total accumulated energy during crack extension is partitioned into the constant cohesive energy and the energy consumed by plastic deformation. It is seen that in the case of high plastic deformation the J-integral at initiation of crack extension can be augmented beyond Γ_0. It can also be seen that the amount of plastic energy increases with the magnitude of the cohesive stress. The reason for this behaviour is that a high value of T_0 requires a high applied load to raise the local stresses to the value of T_0, resulting in a larger plastic zone. A quantitative analysis based on 2D plane strain simulations of C(T) specimens made of 22 NiMoCr 3 7 [1] is shown in Fig. 5.1

K.-H. Schwalbe et al., *Guidelines for Applying Cohesive Models to the Damage Behaviour of Engineering Materials and Structures*, SpringerBriefs in Applied Sciences and Technology, DOI: 10.1007/978-3-642-29494-5_5, © The Author(s) 2013

Fig. 5.1 Effect of plasticity on Γ_0/J_i and δ_0/δ_{5i} ratio, respectively, [1]

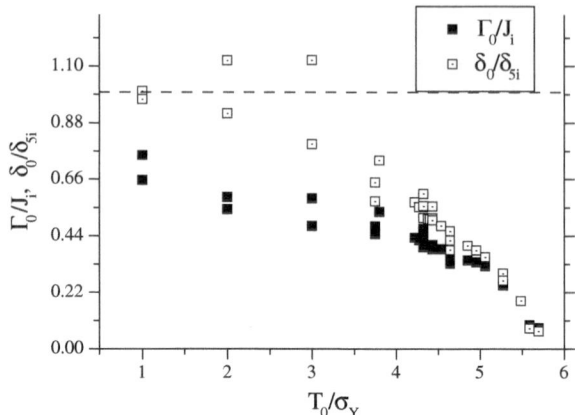

where Γ_0 normalised by J_i is plotted versus T_0 normalised by the material's yield strength, σ_Y. The crack initiation is taken as the point where the first integration point of the cohesive elements has failed. This analysis shows clearly that the assumed equality between Γ_0 and J_i is limited to small scale yielding. In addition, the dependence of the ratio δ_0/δ_{5i} on cohesive strength is shown to be very similar in Fig. 5.1, i.e. for moderate T_0/σ_Y values, the initiation value δ_{5i} can be used for a first estimate of the critical separation δ_0 as well, in case a CTOD R-curve is available instead of a J R-curve.

It should be borne in mind that the results shown in Fig. 5.1 were obtained on a high toughness material whose initiation of crack extension takes place under substantial plastic deformation of the test piece. A high-strength low-toughness material will be much closer to the assumed equality of Γ_0 and J_i.

A further example is given in Fig. 5.2 which shows numerical results of the ratio of plastic work rate to Γ_0 obtained on a C(T) and M(T) specimen as functions of crack extension. It is clearly seen that by far most of the crack extension resistance as measured using classical fracture mechanics methods is plastic deformation, the magnitude of which is strongly influenced by constraint, see Sect. 5.3.

However, it should be pointed out that setting Γ_0 equal to J_i does not necessarily cause substantial errors, simply due to what has been said above, namely, that the total energy absorbed in ductile failure is much greater than J_i so that an inaccuracy in Γ_0 does not substantially affect the result of the simulation of a crack extension resistance curve, see Fig. 3.6. Thus, the pragmatic view of setting Γ_0 equal to J_i is justified.

It is commonly found that different parameter sets determined from numerical optimisations are able to fit experimental results. These parameters may then not have a reasonable physical background. It is, therefore recommended in the present Procedure to use directly determined parameter values as starting values for the numerical optimisation. At least for the cohesive energy this is not feasible when interface cracks have to be treated. Ab initio analyses could then serve for determining physically relevant values.

Fig. 5.2 Ratio of plastic
work rate, dU_{pl}/dA and Γ_0 for
C(T) and M(T) specimens
determined from numerical
simulation of tests [2]

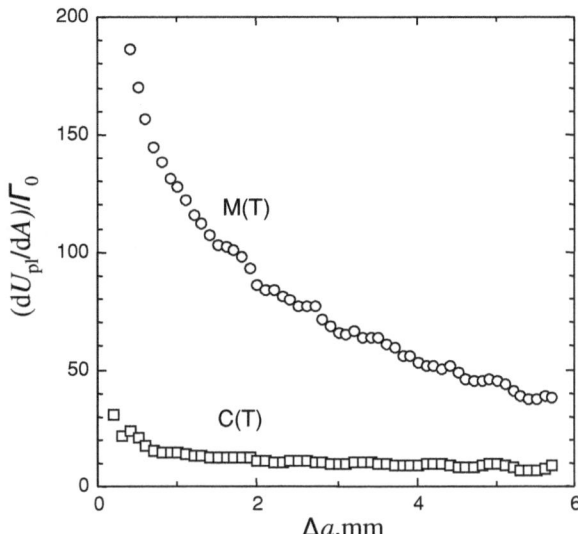

5.2 Shape of the TSL

For the shape of the TSL numerous proposals have been made, as already explained in Sect. 2.2.1, and it has also been pointed out there that different TSL's need different magnitudes of the cohesive parameters. This means that it is essential to work with one TSL for the characterisation of a material, in particular if effects such as triaxiality or mixed mode loading on damage simulation are to be studied.

On a more fundamental basis, the TSL for ductile damage should be derived from the micromechanical observation that the failure mechanism is mainly void nucleation, growth and coalescence. Therefore, the shape of the TSL for ductile damage can be derived from micro-mechanical modelling of a voided unit cell. Such a cell can be modelled either by an axisymmetric model with elastic–plastic material containing a void or by a single element with a Gurson type material model. Under loading with a constant triaxiality a so-called cohesion–decohesion curve [3] can be generated, which relates the elongation of the cell (normalised by the cell height, D) to the mesoscopic (average) stress in the main loading direction, see Fig. 5.3. Such a curve can be directly used as a traction–separation law, or can be approximated by a functional shape, which might be implemented in a finite element code. This example is further elaborated in Sect. 5.3 where the effect of triaxiality is discussed.

In a study on a side-grooved C(T) specimen made of a reactor pressure vessel steel, it was tried to reproduce the experimentally observed crack advance near the specimen surface and near the centre plane. It turned out that the experimentally visible crack front shape can be reproduced with constant cohesive parameters, as will be shown in Appendix 1. However, the parameters depend on the shape of the traction–separation law. The values for three different shapes are given in Table 5.1. The crack front shapes identified by simulations with the different TSLs

Fig. 5.3 Stress-elongation
curve of a voided unit cell
serving as traction–separation
law for the cohesive model

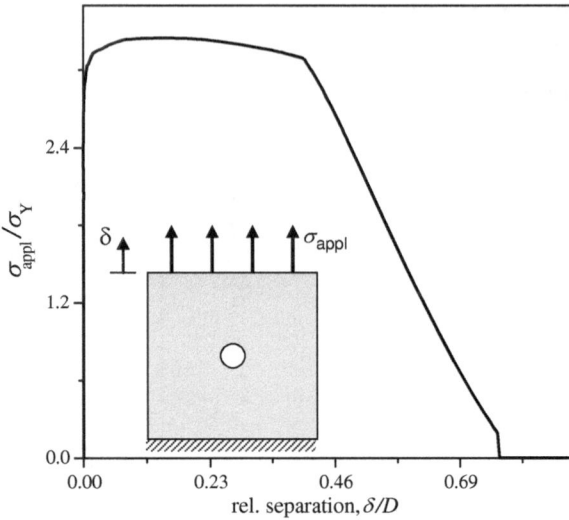

Table 5.1 Parameters optimised for three TSL's to simulate a δ_5 R-curve of the RPV steel
20 MnMoNi 5 5 [4]

	T_0 [MPa]	δ_0 [mm]	Γ_0 [kJ/m^2]
Partly constant	1,500	0.0765	100
Polynomial	1,800	0.0820	83
Cubic decreasing	1,700	0.0926	80

The TSL shapes are shown in Fig. 2.5a

are also different, but all TSL's are able to reproduce the general characteristics of
the experimental shape, thus the optimal shape of the TSL cannot be derived from
these results, see Fig. 5.4. Interestingly, the TSL parameter set of Table 5.1, fits
the global behaviour of a C(T) and an M(T) specimen, Fig. 5.5, i.e. the parameters
can be transferred from one specimen geometry to another independent of the
shape of the TSL.

5.3 Effect of Triaxiality

The stress state is of paramount importance in the fracture behaviour of structural
components. It is here where we observe limits of the transferability of material
properties as obtained from tests using classical fracture mechanics, because the
standard test methods are aimed at determining properties mainly under plane
strain conditions, albeit more recent test methods are designed for plane stress
conditions. Both of these conditions represent extreme stress states between which
a component may be located. It is here where the numerical damage models
promise remedy, however, at least the model presented here still has limits in

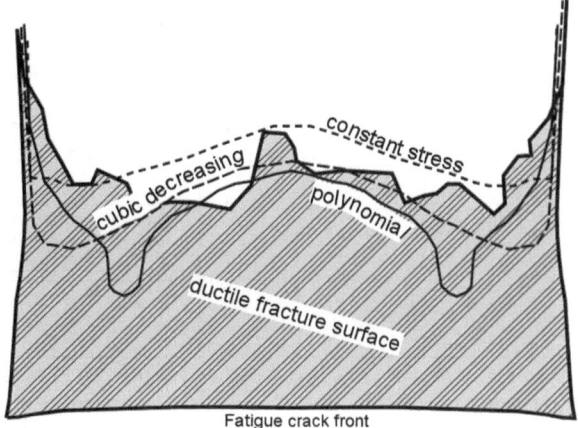

Fig. 5.4 Crack front development in a C(T) specimen of 20 MnMoNi 55 RPV steel with a thickness of 10 mm and 20 % side grooves: Experimental result and simulations using the TSL's from Table 2.1 and the cohesive parameters from Table 5.1 [4]

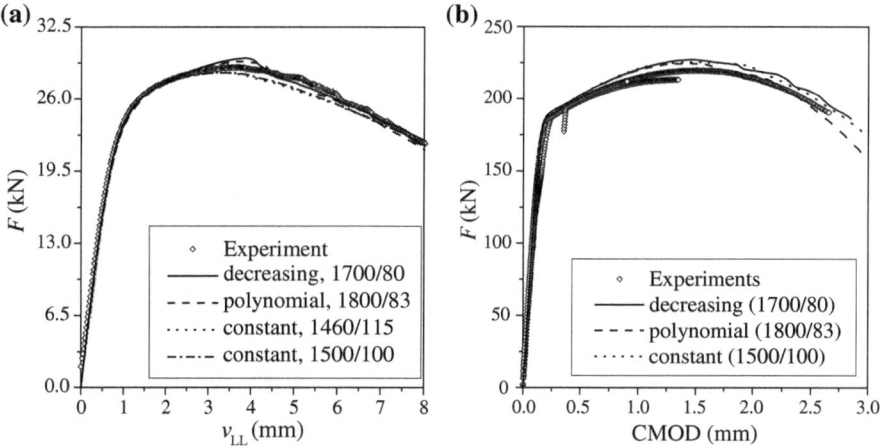

Fig. 5.5 Simulation of fracture specimens made of 20 MnMoNi 55 RPV steel using various traction–separation laws. **a** C(T) specimen; **b** M(T) specimen

transferring parameters as obtained from experiments on simple geometries to an actual component.

The **shape of the TSL** and the magnitudes of the cohesive parameters depend on the stress triaxiality present in the part to be analysed. This is demonstrated by the kind of analysis outlined in Sect. 5.2, namely the derivation of a TSL from a unit cell simulation using the Gurson—Tvergaard—Needleman model [5]. Figure 5.6a shows shapes of TSL's for various degrees of triaxiality, expressed as

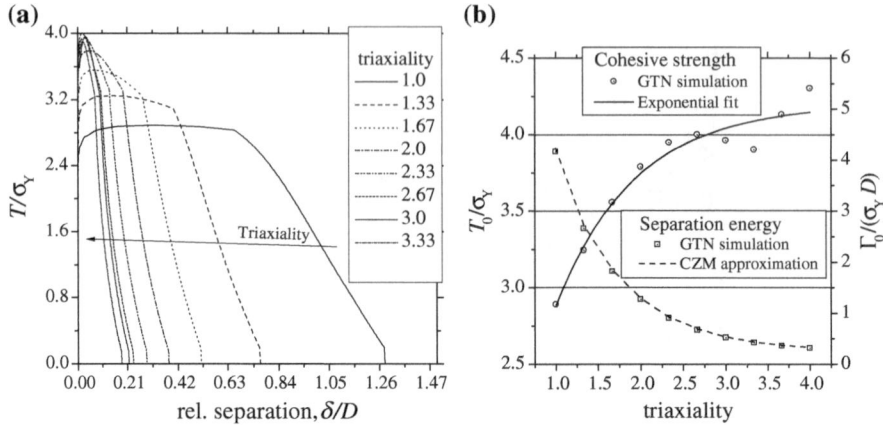

Fig. 5.6 **a** Stress–displacement diagram of a unit cell GTN simulation under various triaxialities; **b** Resulting cohesive strength and cohesive energy as functions of triaxiality [6]. The symbol D denotes the diameter of the unit cell used for determining the cohesive parameters

$h = \sigma_m/\sigma_{eff}$, whereas in Fig. 5.6b the values of T_0 and Γ_0 are shown as functions of h. These can be expressed in analytical form as

$$\frac{T_0}{\sigma_Y} = C_0 - C_1 \exp\left(\frac{h_0 - h}{h_1}\right) \tag{5.1}$$

for the cohesive strength, and

$$\frac{\Gamma_0}{\sigma_Y D} = C_2 - C_3 \exp\left(\frac{h_2 - h}{h_3}\right) \tag{5.2}$$

for the cohesive energy. Instead of two parameters, T_0 and Γ_0, the cohesive model contains now eight parameters C_0, C_1, C_2, C_3, h_0, h_1, h_2, h_3.

For **thin-walled structures** it can be shown that any given TSL with its related cohesive parameters fitted to the δ_5 R-curve of a C(T) specimen made of the aluminium alloy 5083 *T*321, see Fig. 2.4, is able to model the δ_5 R-curve of an M(T) specimen which is different from that of the C(T) specimen [4]. Both specimen types have the same out-of-plane constraint; however, their in-plane-constraint is different; their R-curves are also different, as demonstrated in Fig. 5.7. It is to be concluded that the difference in crack resistance is not due to different cohesive parameters, it follows from the different behaviour of the two specimen types, see Fig. 5.2.

From the same alloy flat notched tensile specimens were investigated. Their fracture behaviour was simulated using two of the TSL's from Table 2.1 and Fig. 2.4. According to Fig. 5.8 the simulations miss the actual behaviour by large margins. In the tensile specimens the maximum stress triaxiality in the range 0.35–0.55 is substantially lower than that of the cracked specimens, having a triaxiality of approx. 0.6–0.65. The low thickness of both specimen types may be

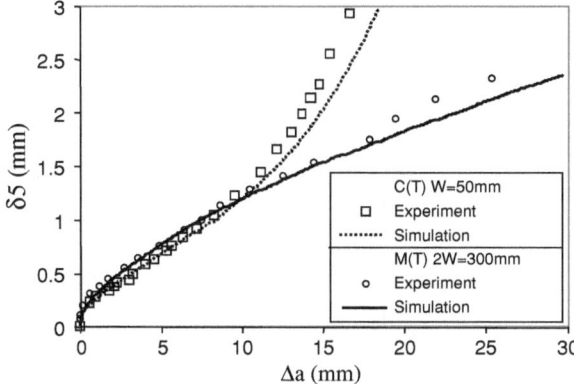

Fig. 5.7 Diagram showing experimental and simulated CTOD R-curves of Al5083 T321 metal sheet (thickness 3 mm) [4]. The C(T) specimen was used for parameter identification, the simulation of the M(T) specimen is a prediction

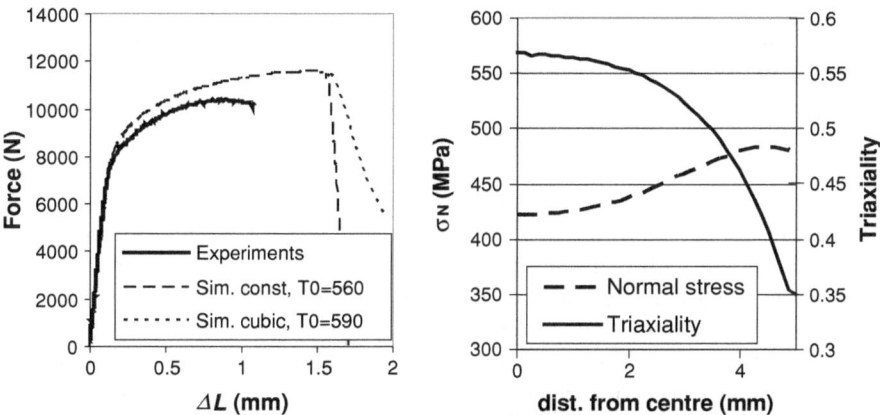

Fig. 5.8 Notched flat tensile specimens with a notch root radius of R = 4 mm, made of Al5083 T321 metal sheet, comparison of experimental tensile test with simulations using the TSL's in Fig. 2.4a and e [4]

misleading: Three-dimensional finite element analyses of cracked specimens [7] show that even in 1 mm thin specimens the stress state in the centre plane is plane strain, thus giving rise to a high T_0 value which can not be achieved in the notched flat tensile specimen. However, if only precracked specimens are investigated, the parameters identified with coupon specimens can be transferred to the structure if both the specimen and the structure have the same out-of-plane constraint, which is usually the case if the thickness is equal. Care must be taken if the thicknesses are not equal; it turned out in [8] that 3D simulations exaggerated the constraint effect of specimens with different thickness $t = 1.9$, 2.9 and 4.15 mm,

Fig. 5.9 Variation of the cohesive parameters across the thickness of a C(T) specimen made of a low strength steel [9]

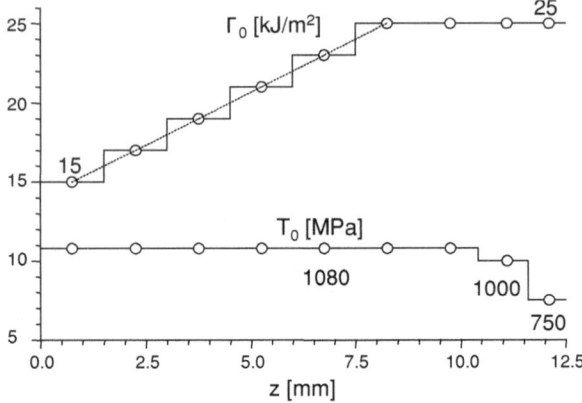

which are plane stress globally, and yielded different R curves whereas in the experiment the CTOD R curves were equal.

In **thick-walled structures** constraint varies through the wall thickness, and the conditions vary with the magnitude of the thickness. Hence, components with different thicknesses exhibit different out-of-plane constraint which, in addition, varies along the thickness direction. Variable constraint across the thickness has consequences for the behaviour of the crack front in the interior of the wall, which the cohesive model should be able to reflect. This topic was already mentioned in Sect. 5.2 showing a crack front of a side-grooved C(T) specimen made of a reactor pressure vessel steel, which was to be reproduced by the cohesive model. It turned out that the general characteristics of the experimental shape could be reproduced, cf. Fig. 5.4, even with constant values for cohesive strength and energy. Thus it can be concluded that the cohesive model is able to consider the different stress states ahead of the crack tip with reasonable accuracy. According to Fig. 5.5 this is due to the high constraint, at which any constraint differences do not lead to strongly varying cohesive parameters, and thus one TSL parameter set models the global as well as the local behaviour of a C(T) specimen.

In contrast to the study just mentioned, a variation of the cohesive parameters across the thickness of a C(T) specimen made of a low strength steel is shown in Fig. 5.9 [9]. In this study, the cohesive parameters were obtained by varying T_0 and Γ_0 such that the simulated crack front development matched the experimentally observed behaviour. As expected, the cohesive strength is highest in the interior and decreases towards the surface of the specimen. The opposite trend is observed for the cohesive energy, but less pronounced. However, another study on Al 6082 O, a rather soft but high hardening aluminium alloy, was performed using the same procedure [10], but in this case the crack front shape could not be reproduced with reasonable accuracy.

The above examples show that the constraint problem is rather complex and needs further work. A pragmatic method has to be pursued in that the cohesive parameters have to be determined on specimens whose constraint conditions resemble those of the component to be analysed. Thus, the dream of having a model that can handle all transferability problems with a unique set of parameters has not yet come true. However, for the sake of a pragmatic application of the cohesive model, some general recommendations can be derived from the present state of knowledge:

In thin-walled structures, triaxiality at a crack tip is independent of the loading mode (tension versus bending). Cohesive parameters can be determined on C(T) specimens and transferred to any configuration with the same thickness.

Thick-walled structures exhibit a pronounced triaxiality effect as a function of the loading mode. Hence, the model parameters should be determined appropriately, i.e. for the thickness and loading mode (tension versus bending) under consideration as average values.

Initially un-cracked structures exhibit low triaxiality, and parameters determined on cracked specimens are not suitable, and vice versa. Therefore, in that case the cohesive parameters have to be determined on low constraint specimens.

References

1. Xu, Y., Yuan, H.: Computational analysis of mixed-mode fatigue crack growth in quasi-brittle materials using extended finite element methods. Eng. Fract. Mech. **76**, 165–181 (2009)
2. Brocks, W.: Ductile crack extension in thin-walled structures. In: Advanced School on Nonlinear Fracture Mechanics Models. Udine (I), 14–18 July 2008
3. Brocks, W.: Cohesive strength and separation energy as characteristic parameters of fracture toughness and their relation to micromechanics. SID **1**, 233–243 (2005)
4. Scheider, I., Brocks, W.: Effect of the cohesive law and triaxiality dependence of cohesive parameters in ductile tearing. In: Gdoutos (ed.) Proceedings of the XVI European Conference on Fracture, Alexandropolis, Greece (2006)
5. Broberg, K.B.: The cell model of materials. Comput. Mech. **19**, 447–452 (1997)
6. Scheider, I.; Huber, N.; Schwalbe, K.-H.: Applicability of the cohesive model to engineering problems: On parameter identification for ductile materials. In: Pokluda, J., Lukáš, P. (eds.) Proceedings of the 17th European Conference on Fracture, pp. 2063–2071 (2008)
7. Newman Jr, J., Bigelow, C.A., Shivakumar, K.N.: Three-dimensional elastic-plastic finite-element analyses of constraint variations in cracked bodies. Eng. Fract. Mech. **46**, 1–13 (1993)
8. Scheider, I., Uz, V., Huber, N.: Applicability of the cohesive model to fracture of light-weight structures: parameter identification and thickness dependence. In: 18th European Conference on Fracture, 31 Aug to 3 Sept 2010, Dresden, Germany (2010)
9. Chen, C.R., Kolednik, O., Scheider, I., Siegmund, T., Tatschl, A., Fischer, F.D.: On the determination of the cohesive zone parameters fort he modelling of micro-ductile crack growth in thick specimens. Int. J. Fract. **120**, 417–536 (2003)
10. Hachez, F.: Experimental and numerical investigation of the thickness effect in the ductile tearing of thin metallic plates. PhD thesis, Université Catholique de Louvain (2008)

Appendix A
Worked Examples for the Simulation of Crack Extension

In this section, the simulation of crack extension will be demonstrated for several specimen configurations and materials which have been well documented within the GKSS research activities. These materials are a low and a high strength aluminium alloy and a pressure vessel steel.

A.1 General Comments

When simulating crack extension in a structural component, three major steps have to be performed:

- Creation of the finite element mesh of the component;
- Determination of the basic material properties;
- Decision on the TSL;
- Determination of the cohesive parameters.

It may also be useful to recall that the magnitude of resistance to crack extension depends substantially on the cohesive strength, T_0, whereas the magnitude of the cohesive energy is less effective. The schematic in Fig. 3.5 is substantiated by simulations using the cohesive model depicted in Fig. A.1. These simulations were performed for small scale yielding. The two diagrams show that the resistance to crack extension is also affected by the strain hardening exponent of the material.

Since when using direct procedures for determining the cohesive parameters, for the determination of J_i a fracture mechanics test including crack extension has to be done anyway it is highly recommended to use the test data for validating the cohesive parameters.

K.-H. Schwalbe et al., *Guidelines for Applying Cohesive Models to the Damage Behaviour of Engineering Materials and Structures*, SpringerBriefs in Applied Sciences and Technology, DOI: 10.1007/978-3-642-29494-5, © The Author(s) 2013

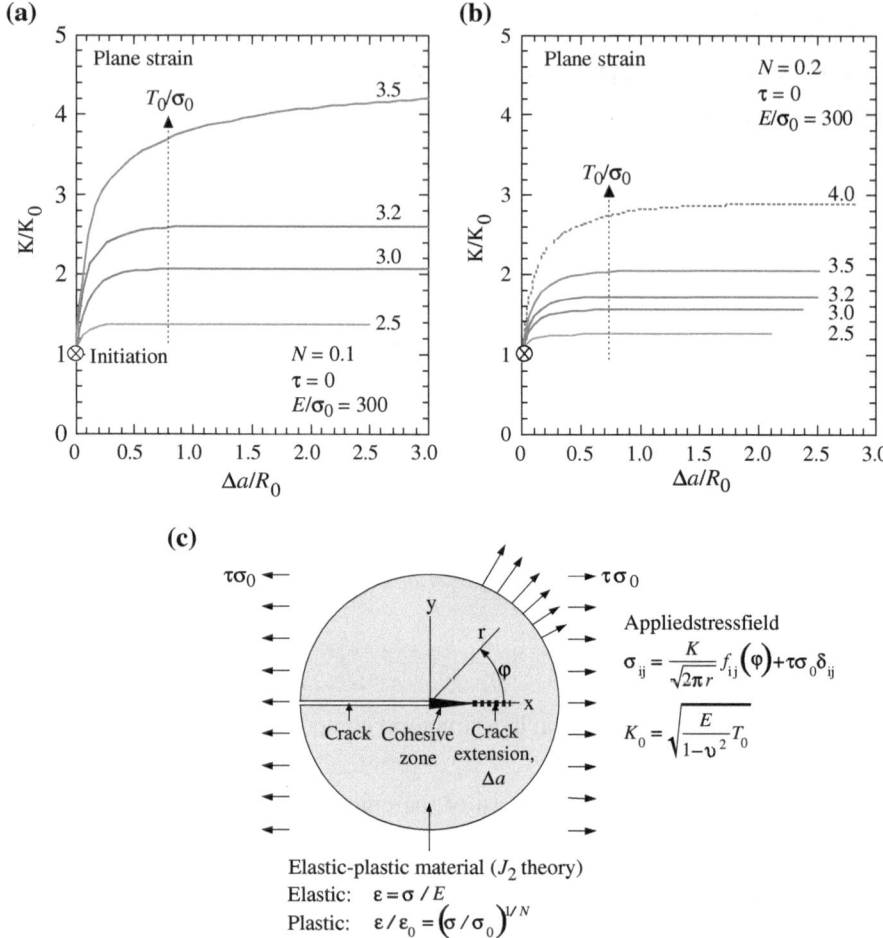

Fig. A.1 Effect of cohesive strength on the resistance to crack extension as determined by cohesive model simulations [1], **a** strain hardening exponent $N = 0.1$, **b** strain hardening exponent $N = 0.2$

A.2 Three-Dimensional Analysis of Crack Extension in the Low-Strength Aluminium Alloy Al 2024-FC Using Direct Procedures

The unusual aluminium alloy 2024-FC has been used at GKSS as a model material for numerous investigations. This material was originally the age hardened high-strength aluminium alloy 2024 T351, which was solution treated and then slowly cooled in the furnace (this is where the acronym FC comes from). The goal of this overaging treatment was to receive a material with a very low yield strength of 81 MPa and a high strain hardening exponent of about $N = 0.3$.

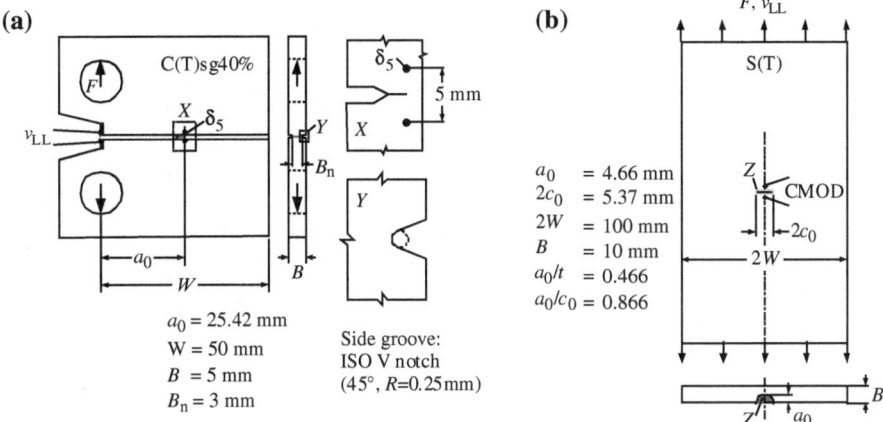

Fig. A.2 Specimen geometries for 3D cohesive model analysis Al 2024-FC [2], **a** side-grooved C(T) specimen, **b** surface cracked tensile panel

A.2.1 Description of Task

In this example [2], structures made of Al 2024-FC were available in the form of 40 % side-grooved C(T) specimens and tensile panels with a surface crack, Fig. A.2. For both specimen types, the force, crack opening displacement, δ_5, the CMOD and the detailed development of the crack front during loading was determined in experiments and then used for validating the cohesive model. In these experiments, the crack front was marked by intermediate unloadings.

The stress–strain curve of the alloy as determined on a round tensile bar is shown in Fig. A.3, the basic material properties are listed in Table A.1.

A.2.2 Cohesive Model

In contrast to the TSL recommended in the present Procedure, this early study employed a rectangular TSL where $T(\delta) = T_0$ and $\Gamma_0 = T_0 \delta_0$. In this TSL, δ_1 is zero, and δ_2 is equal to δ_0, for the meaning of these symbols see Fig. 3.1.

The side-grooved specimens exhibited normal fracture so that for the cohesive strength the method schematically outlined in Fig. 3.2 was applied. The detailed procedure is depicted in Fig. A.4. A sharply notched tensile specimen with a circular cross section was pulled to fracture. The stress distribution over the cross section of the specimen was determined by a axisymmetric stress analysis (with v. Mises plasticity, however, without cohesive elements); the analysis was based on the true stress–strain curve of the material as determined on a smooth tensile specimen as shown in Fig. A.3.

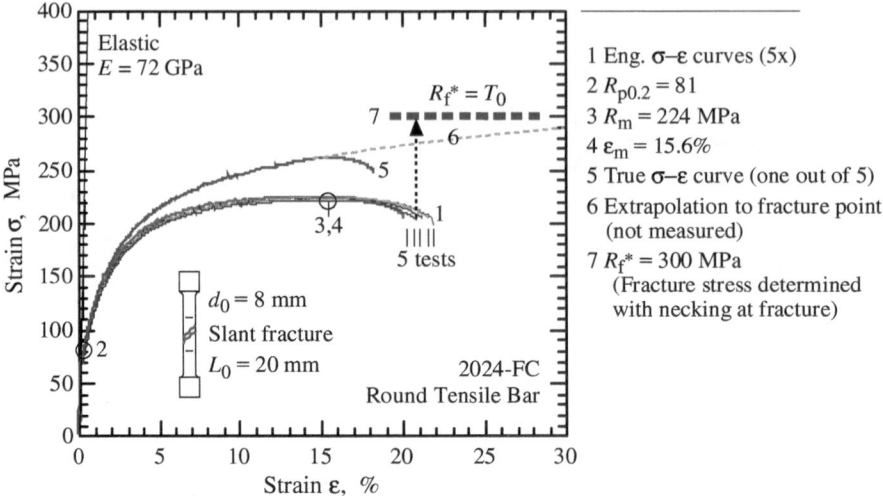

Fig. A.3 Stress–strain curve of Al 2024 FC [2, 3]

Table A.1 Tensile properties of Al 2024 FC

Stress–strain properties	
Elastic modulus	$E = 72{,}000$ MPa
Poisson ratio	$v = 0.3$
Yield strength (at 0.2% plastic strain)	$R_{p0.2} = 81$ MPa
Tensile strength	$R_m = 224$ MPa
True failure strength[a]	$R_f^* = 300$ MPa

[a] This value was determined using the cross section measured after the test

As the diagram shows, at very low applied loads the stress maximum occurs at the notch root, which with increasing load is gradually shifted to the centre of the specimen. At the incident of fracture, the maximum stress amounts to 420 MPa; this is more than five times the value of the yield strength.

Figure A.5 shows the details of the determination of the cohesive energy, Γ_0. As depicted schematically in Fig. 3.4, the line representing the stretch zone width (in this case 10 μm) intersects with the initial section of the R-curve which was determined on five specimens using the electrical potential drop method; reproducibility of these measurements is excellent and in very good agreement with the two calibration points also shown in the diagram. The intersection yields $J_i = 10$ N/mm, which is set equal to the cohesive energy, Γ_0.

NOTE   According to the standard ISO 12135 [4], the J-Δa pairs have to meet the requirements outlined in Fig. 3.4 which have been developed for the multiple specimen method, where each data point is obtained by a test on one specimen. However, in the example shown in Fig. A.5, the single

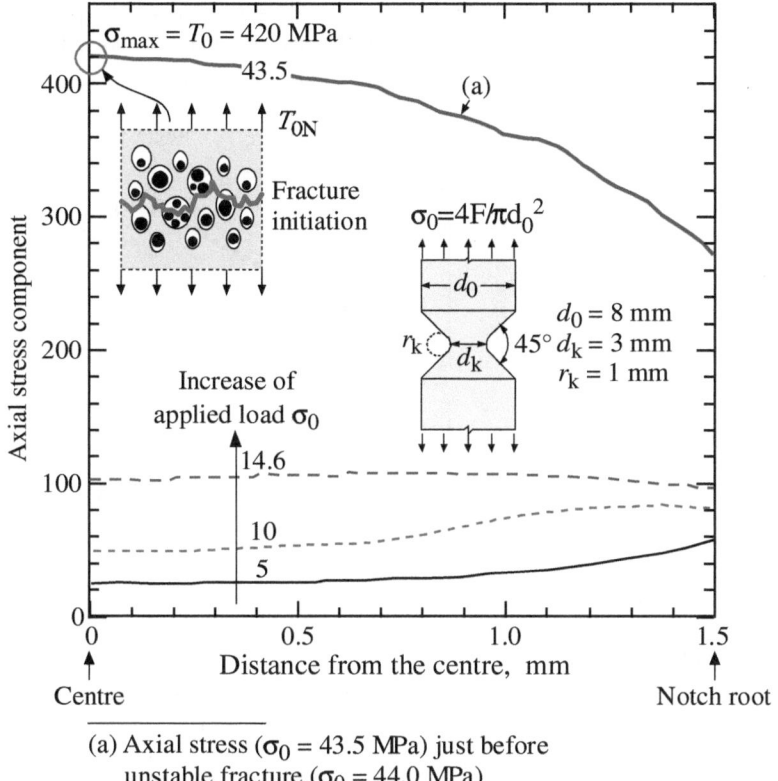

(a) Axial stress ($\sigma_0 = 43.5$ MPa) just before
unstable fracture ($\sigma_0 = 44.0$ MPa)

Fig. A.4 Tensile test and FEM analysis on notched specimen of Al 2024 FC for the determination of the cohesive stress [3]

specimen method was used in the form of the DC potential drop method, although due to their limited accuracy, the standard excludes single specimen methods from the determination of initiation of crack extension. The excellent coincidence of the calibration points in Fig. A.5 with the DC PD data justifies the procedure chosen.

A.2.3 Finite Element Model

A finite element user code [5] was employed. Crack extension simulations were performed using a three-dimensional FE model, i.e. plane cohesive elements were embedded in the symmetry plane at the element faces of hexahedral continuum elements. For the side-grooved C(T) specimen, 6 layers of elements per half net section thickness were used; the side groove was modelled as a sharp notch, with only one extra layer for the side groove. The finite element mesh consisted of 1,258 20-node quadratic elements with 14 reduced integration points and 7,081 nodes.

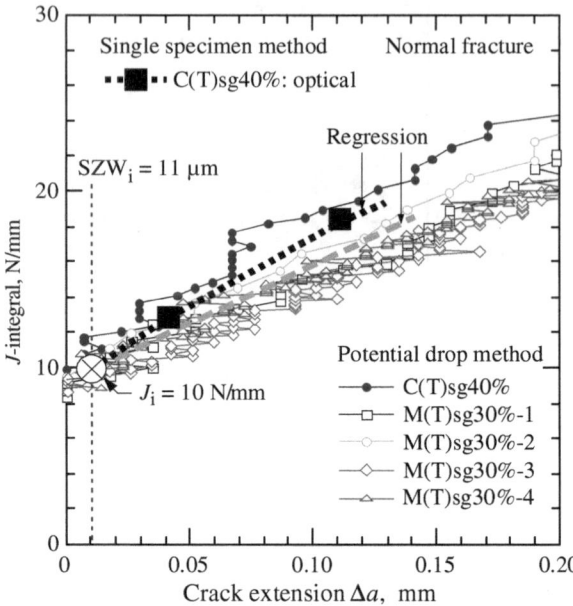

Fig. A.5 Determination of the J-integral at initiation of stable crack extension, J_i, which is the cohesive energy of low strength Al 2024 FC [3]

The near-tip element size was 0.2 mm. In the ligament plane, two-dimensional nine-node quadratic elements were placed. In order to ensure compatibility with nine-node quadratic interface elements, transition elements with 22 nodes were used.

The surface cracked SC(T) specimen was treated similarly to the C(T) geometry. However, the mesh contained 1,645 elements and 9,044 nodes with a smallest element size of 0.25 mm. For the C(T)sg specimen type, also two-dimensional plane strain analyses were performed in order to demonstrate the usefulness of a full 3D analysis.

A.2.4 Analyses and Results

In all the simulations the cohesive parameters were set equal to the values identified above ($T_0 = 420$ MPa, $\Gamma_0 = 10$ N/mm) and considered independent of the stress state.

According to Fig. A.6 the 3D analysis of the C(T) specimen matches the experimental CMOD—crack extension behaviour very well, whereas the plane strain results predict CMOD values which are substantially higher than the experimental ones. The development of the crack front during loading for the same specimen is depicted in Fig. A.7. Due to the presence of the side-grooves, the crack front extends much faster near the specimen surface than in the interior. This is particularly well predicted by the 3D simulation. The surface cracked tensile panel SC(T) was analysed with the 3D FE model only. As can be seen in Fig. A.8, also in this case the simulations match the experiments very well.

Fig. A.6 Crack mouth
opening displacement versus
crack extension for the side-
grooved C(T) specimen

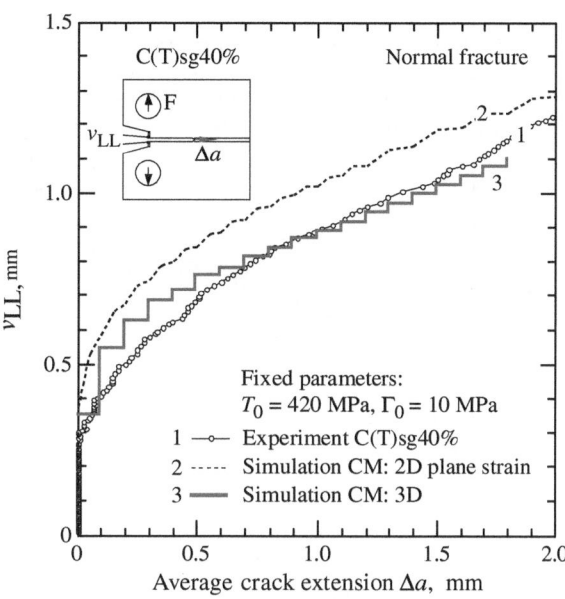

(a) 3D CM simulation of
consecutive crack fronts

(b) Comparison of crack shapes
Experiment vs. simulation

(c) Crack tip region of the 3D mesh

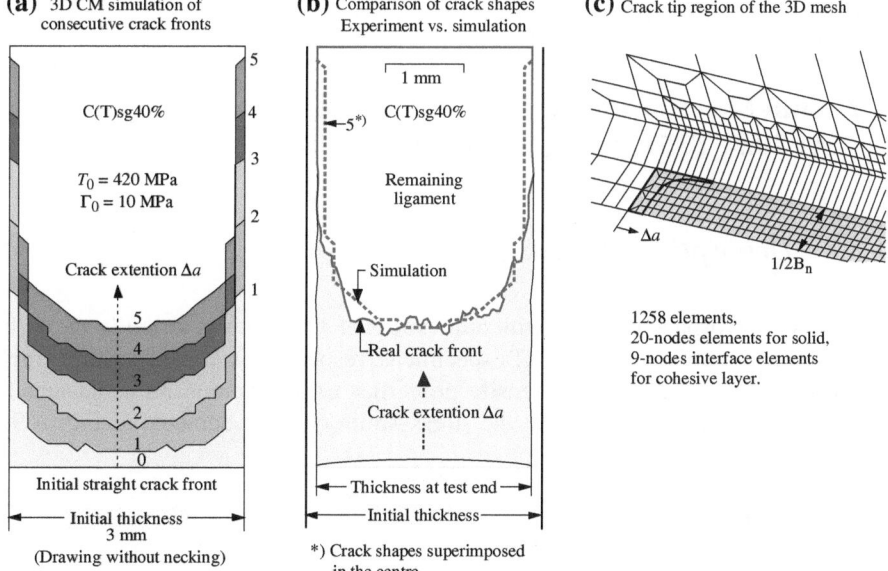

1258 elements,
20-nodes elements for solid,
9-nodes interface elements
for cohesive layer.

Fig. A.7 Side-grooved C(T) specimen showing experimental and simulated crack fronts; **a** crack
front shape development from cohesive simulations, **b** comparison of the crack front from
experiment at the end of interrupted test and respective cohesive simulation, **c** mesh of C(T)
specimen with view on the fracture plane

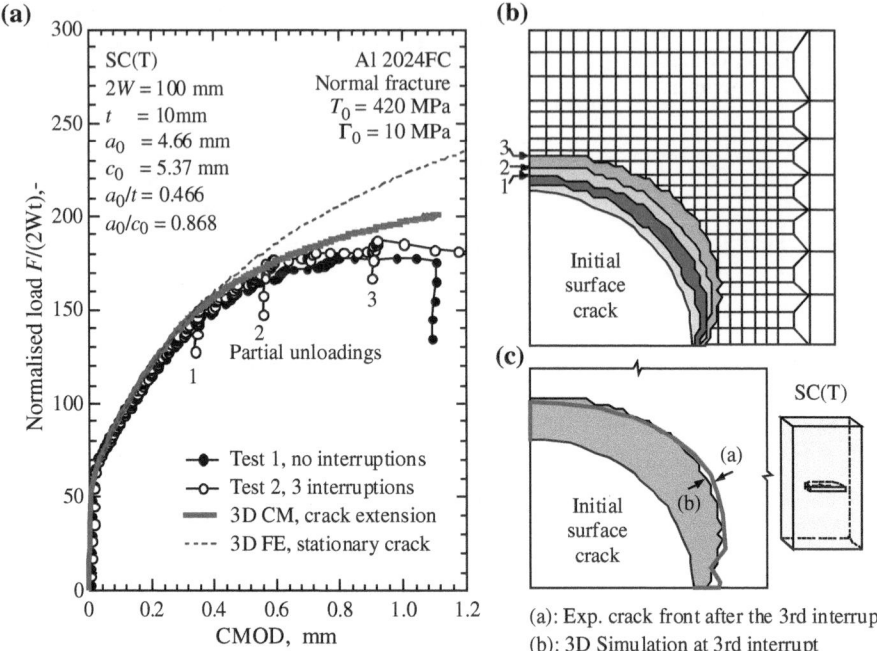

Fig. A.8 Surface cracked tensile panel, **a** applied nominal stress versus the crack mouth opening displacement, **b** development of the crack extension; **c** comparison of simulated and experimental crack shape and extension

A.3 Thin Sheet of the High-Strength Aluminium Alloy Al 2024 T351, Using Direct Procedures

A.3.1 Description of Task

This typical aerospace material with a thickness of 1.6 mm was tested in the form of two 50 and 1,000 mm wide C(T) specimens, respectively, with a starting crack length of $a_0/W = 0.5$ [3]. The tensile properties as determined on a flat tensile specimen are listed in Table A.2; the stress–strain curve is depicted in Fig. A.9.

A.3.2 Cohesive Model

The same rectangular model as in Sect. A.2.2 was utilised. The cohesive strength was determined directly from the stress–strain curve in Fig. A.9. The fracture strength, R_f^*, was obtained from the reduced cross section at fracture similar as in Sect. A.2 for 2024FC; this yielded $R_f^* = T_0 = 550$ MPa. From fracture mechanics

Table A.2 Tensile properties and cohesive parameters of Al 2024 T351	Stress–strain properties	
	Elastic modulus	$E = 72{,}000$ MPa
	Poisson ratio	$v = 0.3$
	Yield strength (at 0.2% plastic strain)	$R_{p0.2} = 300$ MPa
	Tensile strength	$R_m = 427$ MPa
	True failure strength	$R_f^* = 550$ MPa
	Cohesive Parameters	
	J_i from three experiments	8.8–10 N/mm
	Cohesive strength, T_0	550 MPa
	Cohesive energy, Γ_0	9.5 N/mm

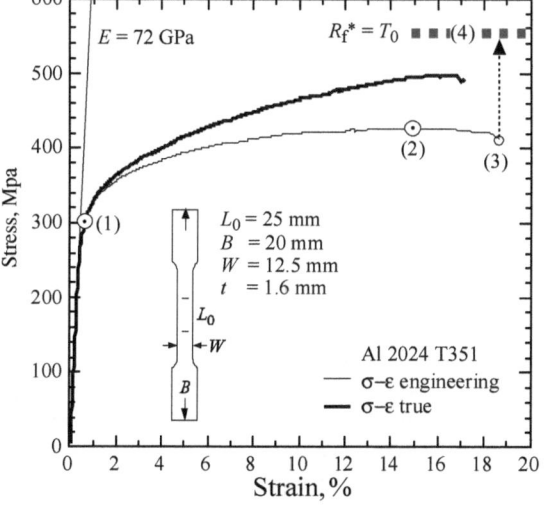

Fig. A.9 Stress–strain curve of Al 2024 T351 [3]

experiments, the *J*-integral at initiation of crack extension, J_i, yielded values between 8.8 and 10 N/mm, Fig. A.10, from which the cohesive energy was set equal to 9.5 N/mm, Table A.2.

A.3.3 Finite Element Model

A finite element user code [5] was employed. The specimens were analysed using a plane stress model consisting of isoparametric 8-node solid elements with the smallest elements near the crack tip having a width of 0.05 and height of 0.075 mm.

Fig. A.10 Determination of the *J*-integral at initiation of stable crack extension, J_i, which represents the cohesive energy of high-strength Al 2024 T351 [3]

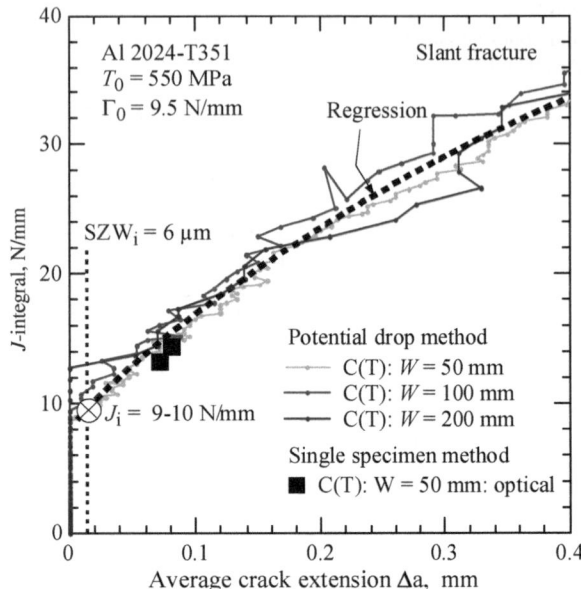

A.3.4 Results

The FE and cohesive models, together with the directly determined cohesive parameters reproduce the experimental crack extension resistance curve of the small specimens very well, Fig. A.11a. The application to the large specimens results in a slight under-prediction of the crack extension resistance for large amounts of Δ*a*, Fig. A.11b. A possible reason for this behaviour may be the occurrence of a small amount of buckling in the larger specimens, resulting in somewhat larger measurements of δ_5.

A.4 Three-Dimensional Analysis of Crack Extension in the Pressure Vessel Steel 20 MnMoNi 55, Using Direct Procedures

A.4.1 Description of Task

For this exercise, two side-grooved C(T) specimens and one side-grooved M(T) specimen, Fig. A.12, were machined from the pressure vessel steel 20MnMoNi55. Crack front development and force—load line displacement curves were determined on the C(T) specimens, whereas on the M(T) specimen the *J*-Δ*a* curve and the force—CMOD curve were investigated [3].

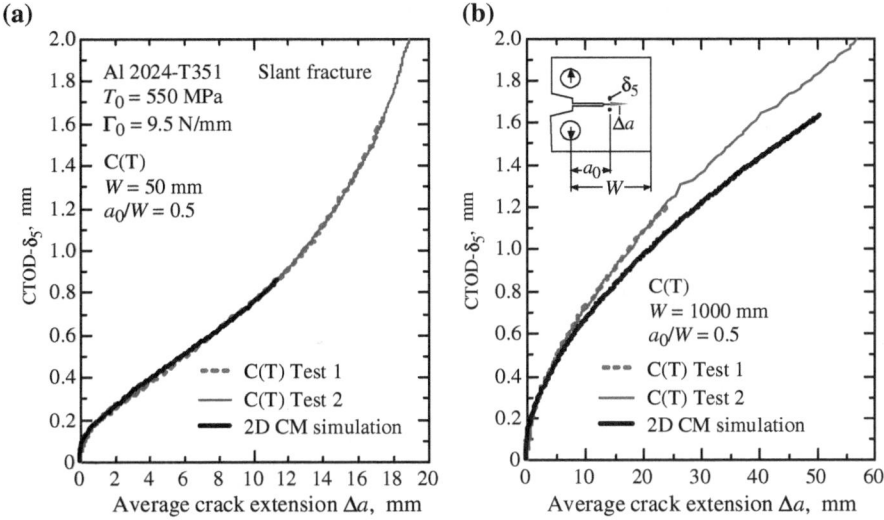

Fig. A.11 Experimental and simulated δ_5–Δa curves of high-strength Al 2024 T351, **a** 50 mm wide C(T) specimens, **b** 1,000 mm wide C(T) specimens

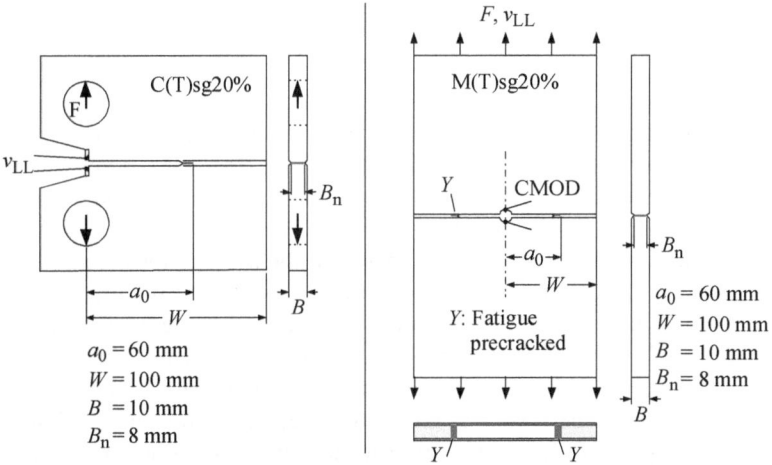

Fig. A.12 Specimen geometries of the steel 20 MnMoNi 55 for 3D cohesive model analysis, **a** side-grooved C(T) specimen, **b** side-grooved M(T) specimen

A.4.2 Cohesive Model

The standard GKSS cohesive model depicted in Fig. 3.1 served for the analysis. For the C(T) specimen, a further simulation was run with the rectangular TSL mentioned in Sect. A.2.2.

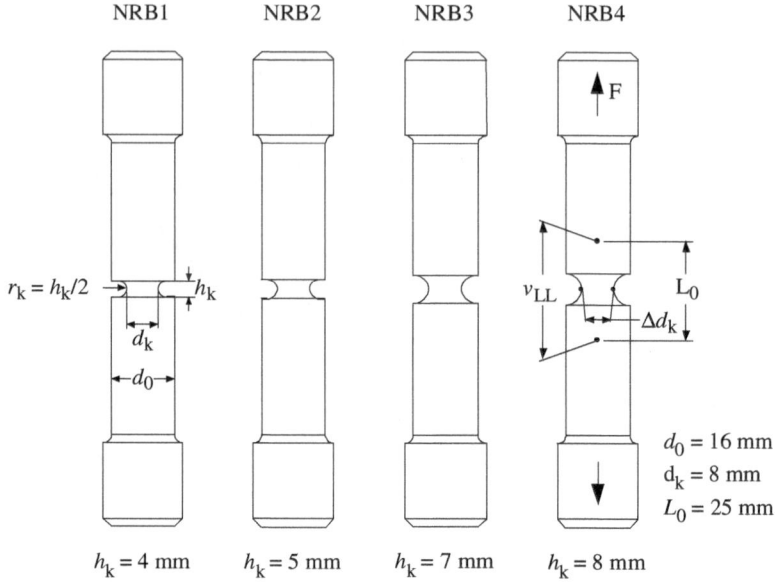

Fig. A.13 Round notched tensile specimens for the determination of the cohesive strength of 20MnMoNi55 steel

Also in this case, only normal separation was considered. For the pressure vessel steel 20MnMoNi55 the cohesive stress was determined by means of a series of round notched tensile specimens, Fig. A.13. The FEM analysis of these specimens was performed as already depicted in Fig. A.4 and is shown in Fig. A.14, resulting in T_0 equal to 1,460 MPa. The result of Fig. A.14 is rearranged and shown again in Fig. A.15 where the average tensile stress, applied force divided by the original net cross section, is compared with the stress in the centre of the specimen as determined by the finite element analysis. The augmentation of the maximum stress in the specimen centre over the average stress due to the notch effect is clearly visible and amounts to about a factor of two at the fracture point. The results obtained on three of the four notched specimens are shown in Fig. A.16; it is interesting to note that the values of T_0 are almost identical and hence independent of the stress triaxiality σ_m/σ_{eff} which varied between 1.053 and 1.489.

Using the same procedure as applied to the aluminium alloy, Fig. A.5, the cohesive energy, Γ_0, was determined with a value of 120 N/mm, Fig. A.17. Again, excellent reproducibility of the experiments carried out on four specimens, with both the electrical potential method and the multiple specimen method can be seen from the two diagrams.

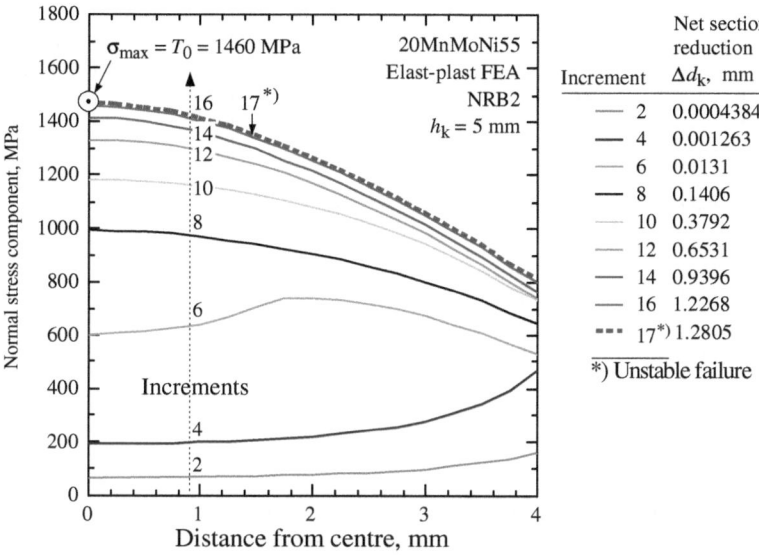

Fig. A.14 Tensile test on 20 MnMoNi 55 steel and FEM analysis for the determination of the cohesive stress using the direct method; presented for specimen NRB2 [3]

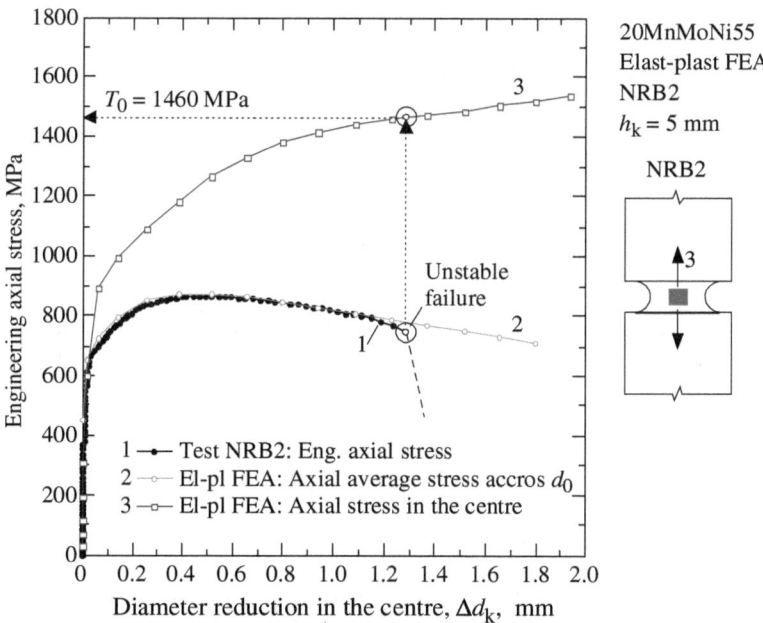

Fig. A.15 Tensile test on 20 MnMoNi 55 steel and FEM analysis for the determination of the cohesive stress using the direct method; development of the axial stresses and use of experimental limit at unstable failure point; presented for specimen NRB2 [3]

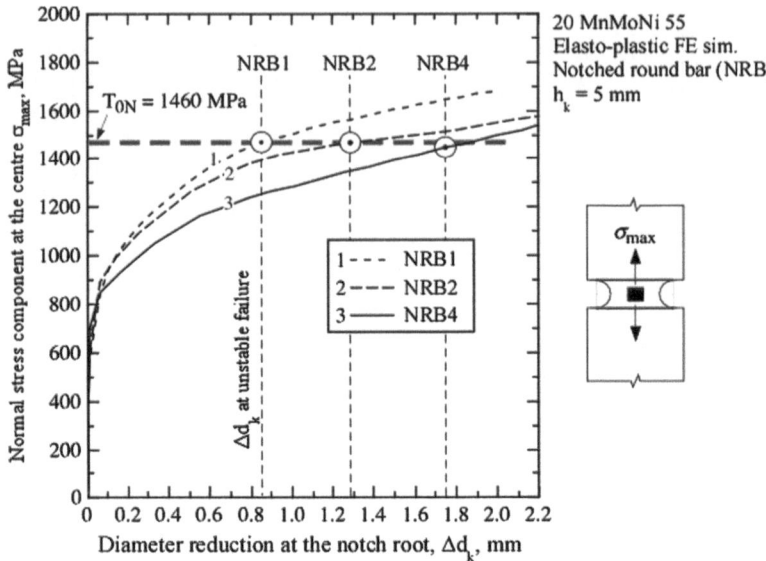

Fig. A.16 Tensile test on 20 MnMoNi 55 steel and FEM analysis for the determination of the cohesive stress using the direct method; development of the axial stresses and use of experimental limit at unstable failure point; for three specimens NRB1, NRB2, NRB4 with different notch sizes

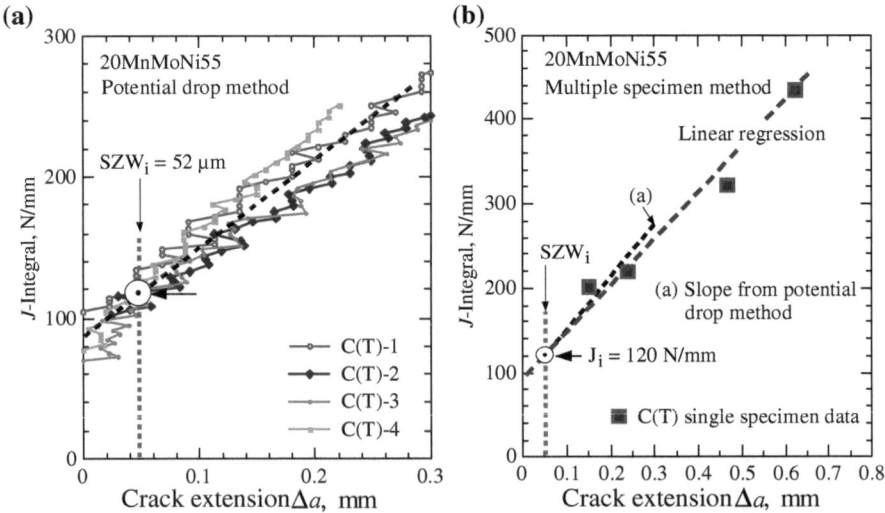

Fig. A.17 Experimental determination of the cohesive energy for normal fracture of the steel 20 MnMoNi 55 via the J-integral at initiation of ductile crack extension [3]

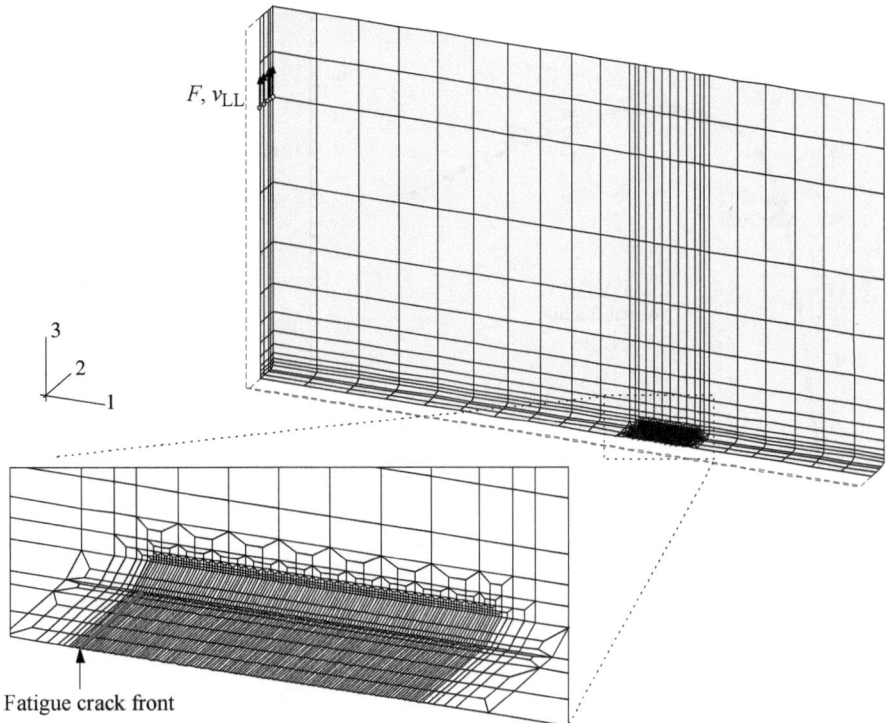

F, v_{LL}

3
2
1

Fatigue crack front

Fig. A.18 Finite element mesh of the side-grooved C(T) specimen with the detailed view on crack tip region and side-grooved notch

A.4.3   Finite Element Model

The simulations were conducted using the finite element code ABAQUS with additional user defined cohesive elements. The 3D mesh of the C(T) specimen modelled one quarter due to symmetry. The application of the load through the bolt was simplified by a single line of nodes where the displacement was applied, see Fig. A.18. The mesh had 7 layers of cohesive elements across the half thickness modelled, allowing for 9 mm of crack extension. The side groove was modelled as a sharp V-notch. In total 6,732 continuum elements and 910 cohesive elements were generated, which yielded 28,700 degrees of freedom. The mesh of the M(T) specimen was very similar, except the boundary conditions (displacements were applied at the top nodes of the mesh and additional boundary conditions to model the threefold symmetry) and additional elements to increase the height of the FE model.

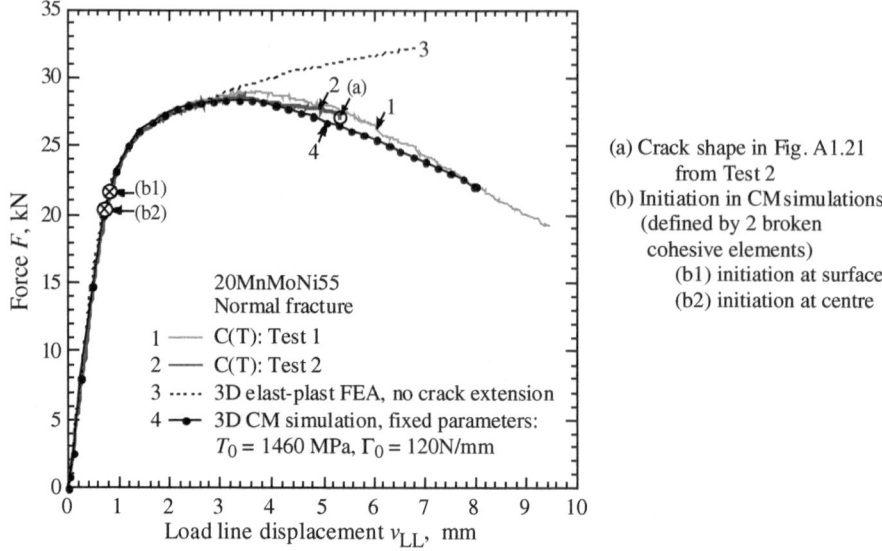

Fig. A.19 Experimental and simulated force–load-line displacement behaviour of two identical side-grooved C(T) specimens of 20 MnMoNi 55

A.4.4 Results

Figures A.19 and A.20a demonstrate the ability of the cohesive model to reproduce with high accuracy the force–displacement behaviour of both specimen geometries. It is particularly interesting that the complete shape of the curve is well captured with a single set of cohesive parameters. This is a strong indication that in the range of pre-cracked structures the assumption of constant cohesive parameters is justified. As in the case of the specimen of the low strength aluminium alloy shown in Fig. A.7, the cohesive model analysis was able to model the crack front shape of a side-grooved C(T) specimen very well, Fig. A.21 (the same situation is also depicted in Fig. 5.4).

A.5 Simulation of Crack Extension in a Stiffened Structure, Using Indirect Identification Procedure

A.5.1 Description of Task

In 2006 the American Society of Testing and Materials organised a predictive round robin exercise with 3 participants. In this round robin, crack extension in an integrally stiffened panel was to be predicted, and the participants were free to use

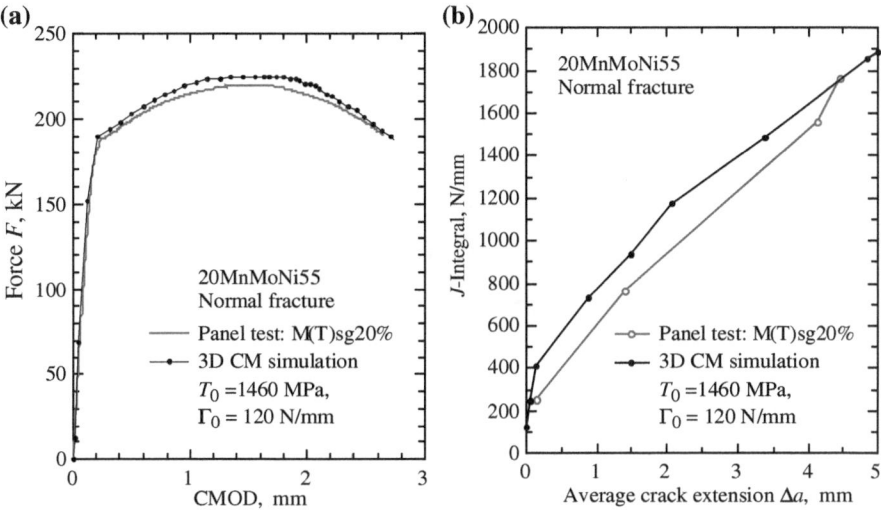

Fig. A.20 Experimental and simulated behaviour of a side-grooved M(T) panel of 20 MnMoNi 55, **a** force–CMOD curve, **b** J-resistance curve

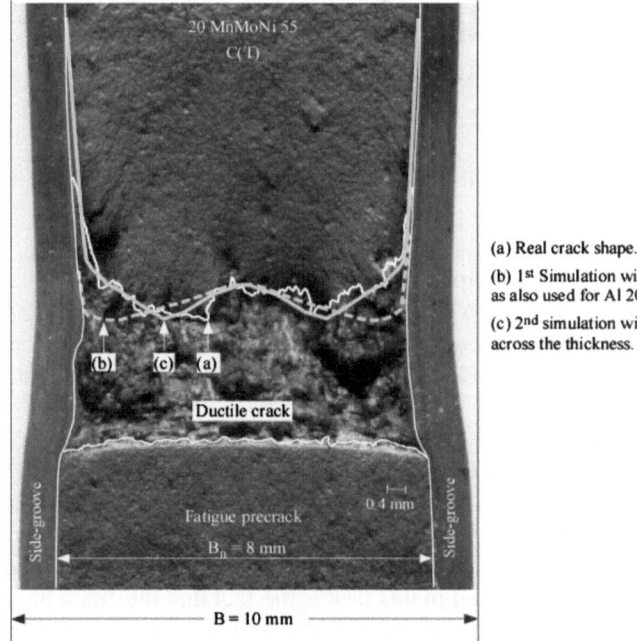

(a) Real crack shape.

(b) 1st Simulation with initial 3D mesh as also used for Al 2024 FC.

(c) 2nd simulation with finer mesh across the thickness.

Fig. A.21 Crack front development in the side-grooved C(T) specimen of 20MnNi55, comparison of experimental and simulated crack front (the end of this test is indicated in Fig. A.19)

Applied remote stress

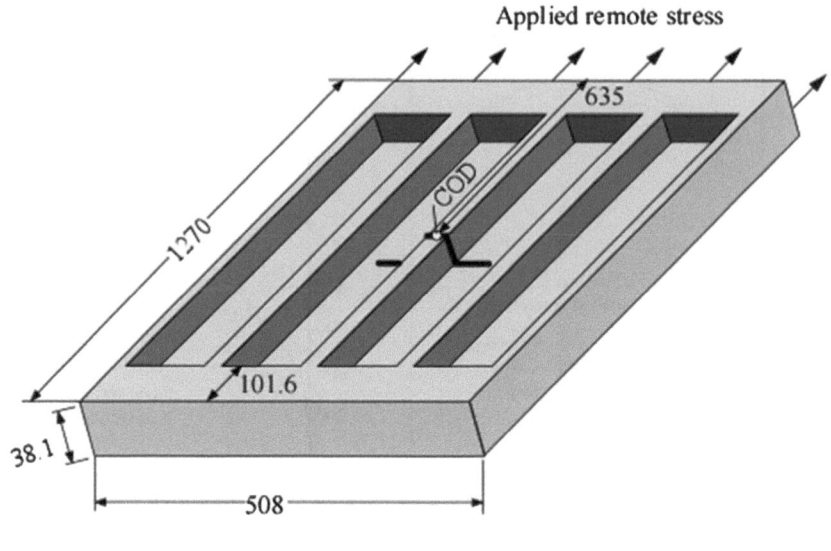

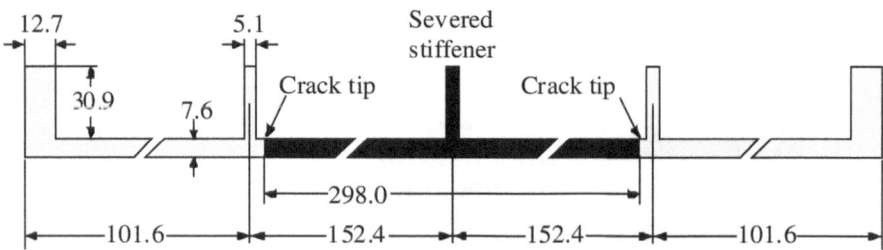

Fig. A.22 Integrally stiffened tensile panel made from Al 2024 T351, drawing provided by Alcoa

their preferred model. The participants were given the drawing of the integrally stiffened panel machined from a 1.5″ thick 2024-T351 plate, Fig. A.22, and provided by Alcoa. The background of this activity was given by the current discussion on using welded-on stiffeners, hence integrally stiffened, components in aerospace structures.

Further information provided was the stress–strain data and a table for force, COD, Δa and K_{eff} of an M(T) panel with the dimensions $W = 406$ mm, $t = 6.44$ mm and $a_0 = 51.5$ mm. The tests to be simulated were performed by Alcoa. From the data of the M(T) panel the cohesive parameters were determined by numerical fitting, and crack extension in the stiffened panel was predicted. The major benefit of the cohesive model in this task is the fact that the crack branching along the skin and into the stiffener is a direct result of the simulation. No assumptions on how the crack grows at the junction are needed.

Table A.3 Parameters identified for Al 2024-T351 from an M(T) specimen using two different finite element meshes

	T_0 MPa	Γ_0 N/mm	δ_0 mm
Shell simulation	770	11	0.018
3D simulation	970	20	0.024

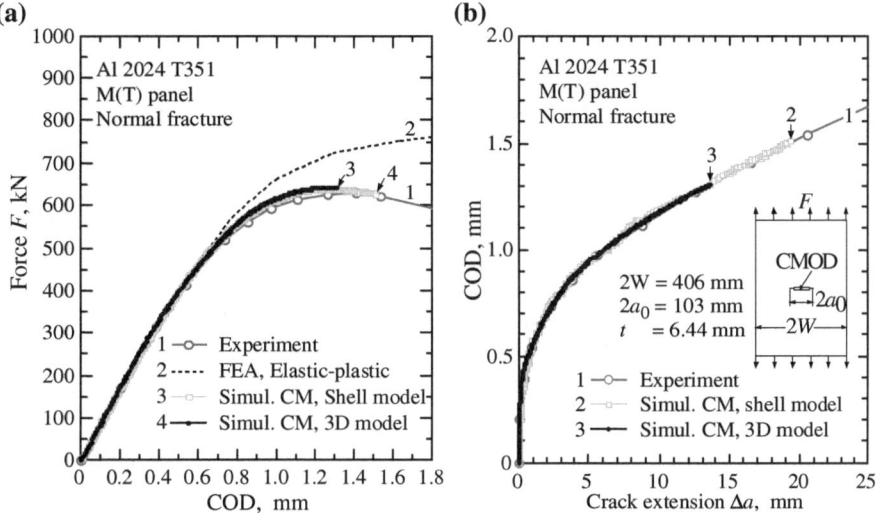

Fig. A.23 Simulations of the M(T) specimen using the optimised parameter sets for the shell and 3D model, **a** force–COD curve, **b** COD–Δa curve [6]

A.5.2 Cohesive Model

The standard GKSS cohesive model depicted in Fig. 3.3 served for the analysis. The parameters were identified by trial and error based on the F(COD) curve and the COD(Δa) curves. Two different finite element models were employed both for the parameter identification and the prediction of the stiffened structure: a shell model and a 3D model. It is crucial that the element types used for the identification are the same as those used for the prediction. The parameters identified for both types are listed in Table A.3. One can see that the values differ significantly, due to the different approximation of the stress state ahead of the crack tip. The results of the simulations of the M(T) specimen with the thus identified parameters are shown in Fig. A.23.

A.5.3 Finite Element Model

Two types of FE models were employed:

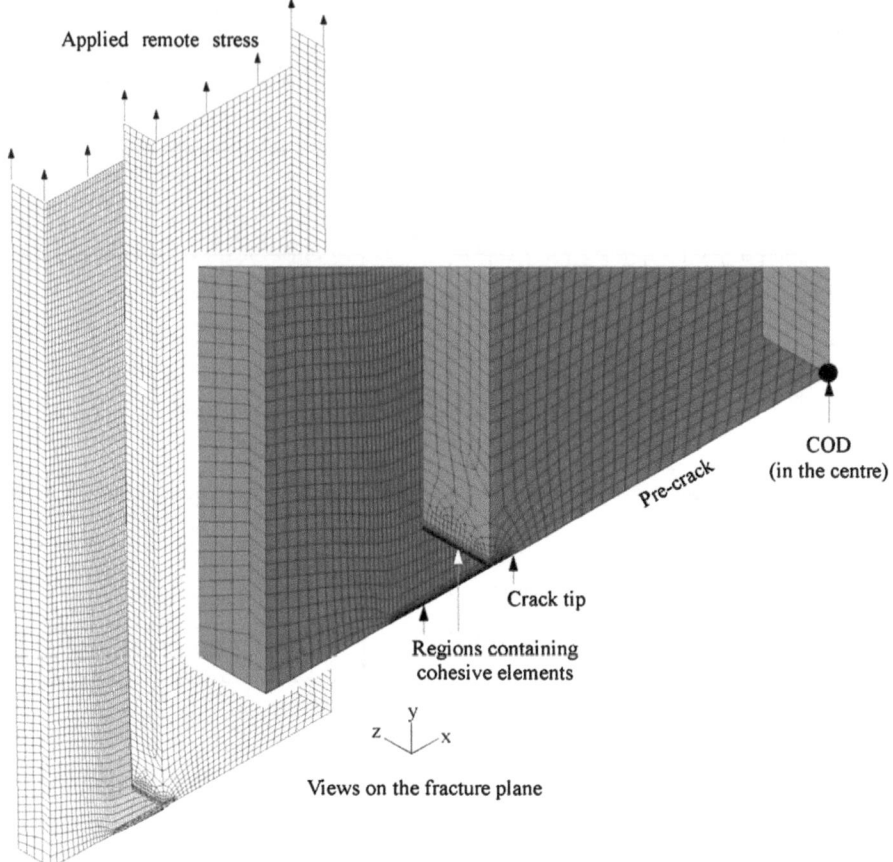

Fig. A.24 Finite element mesh of the stiffened panel using shell elements [6]

(a) Shell model:

Only the mid-plane is used for the shell model of the stiffened panel. This leads to a largely reduced number of elements and nodes (8,782 linear shell elements and 263 cohesive elements, 56,070 DOFs). One quarter of the structure was used because of symmetry. The mesh is shown in Fig. A.24.

(b) 3D model:

Due to the thickness of the panel (7.6 mm for the skin and 5 mm for the stiffener), a large number of element layers in thickness direction are needed, which leads to 41,294 linear continuum elements and 4,625 cohesive elements in the present case, and thus the computation time is much longer, but also the mesh generation is quite complex, see the mesh used for the simulation in Fig. A.25.

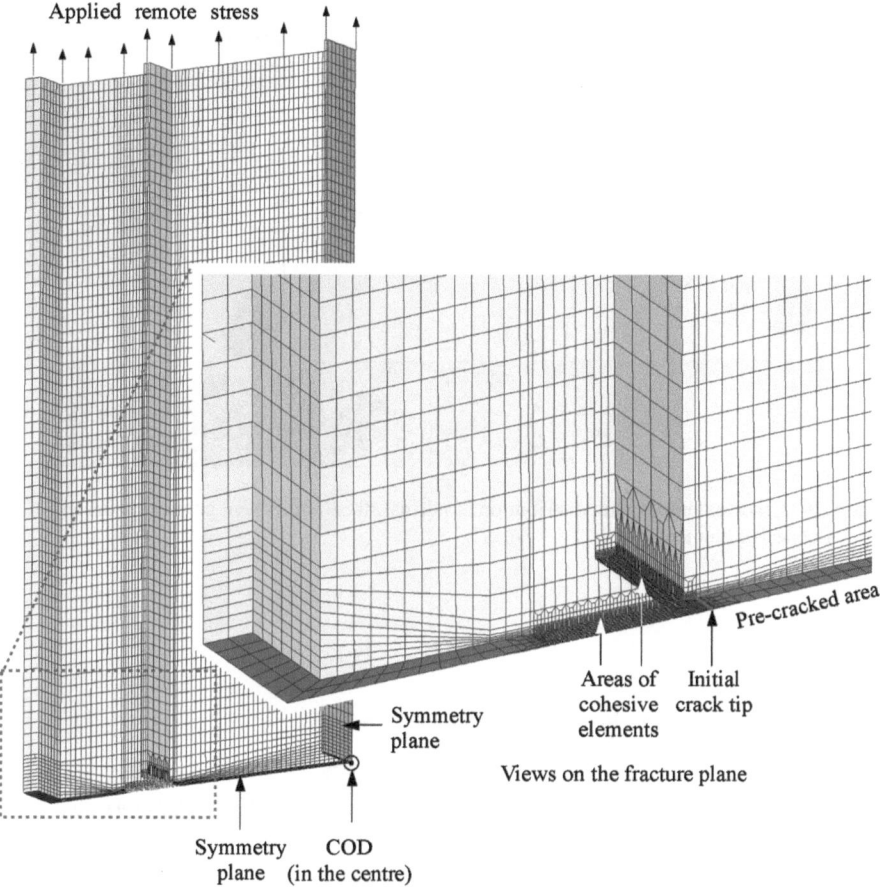

Fig. A.25 Finite element mesh of the stiffened panel using 3D elements [6]

A.5.4 Results

Figure A.26 demonstrates the way the pre-crack propagated through the stiffened panel. When approaching the stiffener, the crack branched and severed the stiffener, and kept on propagating through the skin. This kind of behaviour can be easily modelled with the cohesive model; the only requirement is that cohesive elements have to be placed along the expected crack path. More importantly, the predicted global behaviour should follow as closely as possible the experimental result. This requirement is very well met as demonstrated by Fig. A.27, showing that the shell element model is very close to the experiment; the residual strength is $\sigma_{\text{appl}} = 151$ MPa compared to the actual value of $\sigma_{\text{appl}} = 148$ MPa in the test. The 3D simulation was a bit more conservative leading to $\sigma_{\text{appl}} = 135$ MPa.

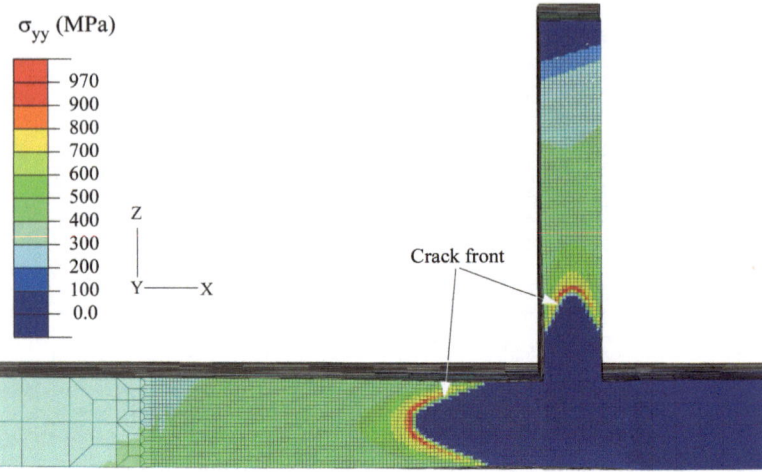

Fig. A.26 Crack extension with distribution of normal stress component at load maximum, 1,400 broken cohesive elements [6]

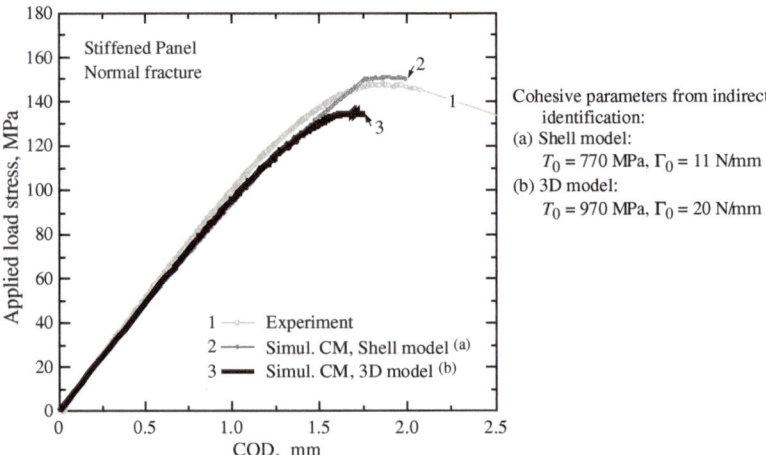

Fig. A.27 Comparison of load-COD curves from experiment and the cohesive simulations of shell and 3D model [6]

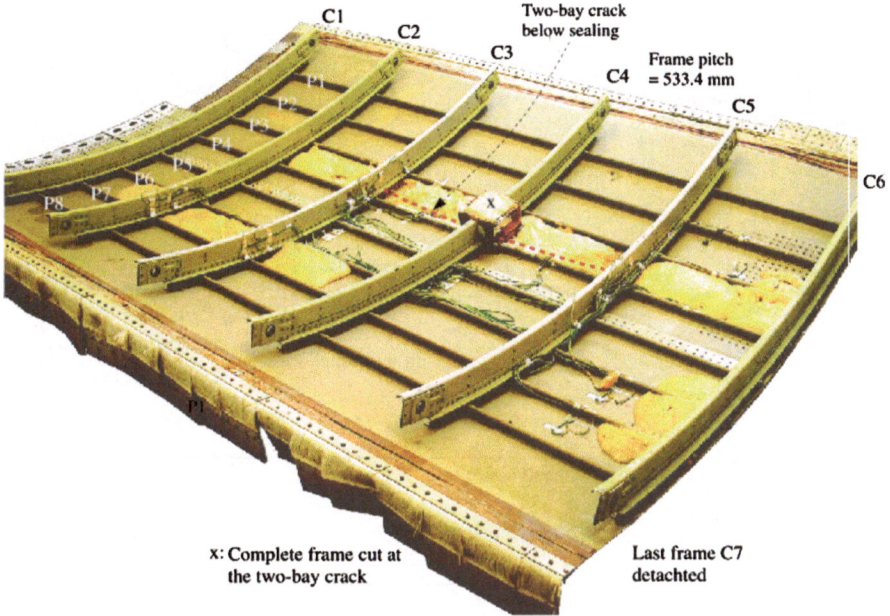

Fig. A.28 Stiffened fuselage panel tested at IMA, Dresden

A.6 Simulation of Residual Strength of a Large Fuselage Panel

A.6.1 Description of Task

The residual strength of a curved riveted fuselage panel, containing a two-bay crack, was tested at IMA (Institut für Materialforschung und Anwendungstechnik GmbH, Dresden, Germany). This test presented a unique opportunity for applying the cohesive model to a complex structural system [7]. Therefore, a comprehensive modeling activity was undertaken in order to simulate the extension of the fatigue crack created in the experiment, with the final goal to estimate the maximum pressure attained in the model and to compare it with the experimental value.

The tested panel is part of a wide-body fuselage with an outer diameter of 5,640 mm, Fig. A.28. The dimensions of the panel are 2,240 by 3,865 mm. The 1.8 mm thick skin is stiffened by seven frames and eight stringers.

As this investigation included numerous details, both for testing and modelling, only the main items can be presented here. For more details see the published paper [7].

Fig. A.29 Test pieces for identifying the cohesive parameters for both normal and slant fractures. The smooth flat bar was also used for the determination of the stress–strain curve; after [7]

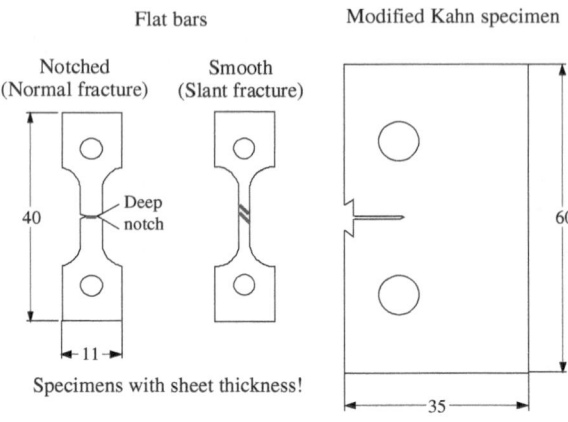

Flat bars   Modified Kahn specimen

Notched   Smooth
(Normal fracture) (Slant fracture)

Specimens with sheet thickness!

Table A.4 Cohesive parameters of the skin material Al 2525 T351

Cohesive strength	$T_{0N} = 660$ MPa
	$T_{0S} = 610$ MPa
Cohesive energy	$\Gamma_{0N} = 40$ N/mm
	$\Gamma_{0S} = 25$ N/mm

A.6.2 Cohesive Model

The standard GKSS cohesive model depicted in Fig. 3.3 served for the analysis. The parameters were identified by a fracture mechanics test. A modified and fatigue pre-cracked Kahn specimen was fabricated from the skin material Al 2525 T351 for the determination of an R-curve, Fig. A.29. During the test, the applied force, F, and the crack mouth opening displacement (CMOD), and were measured.

After a short initial flat portion, the crack propagated in the slant mode as usually observed in thin walled materials. Following this observation, the cohesive parameters were determined for both fracture modes. They are marked with the subscript N for flat fracture and S for slant fracture, and are listed in Table A.4. For flat fracture, the parameters were identified as described in Sect. 3.4.1, using the notched micro tensile specimen shown in Fig. A.29. For slant fracture, the cohesive strength was determined as shown in Sect. A.3.2 using the smooth tensile specimen in Fig. A.29. The cohesive energy for slant fracture was determined by comparing the experimental F-CMOD curves of two modified Kahn specimens with a finite element analysis employing the cohesive model, Fig. A.30. The test on the Kahn specimen was then also used to fine-tune the cohesive strength value T_{0N} which is the one listed in the table.

The tensile properties were measured on the tensile specimen in Fig. A.29; the resulting yield strength amounts to $R_{p0.2} = 276$ MPa.

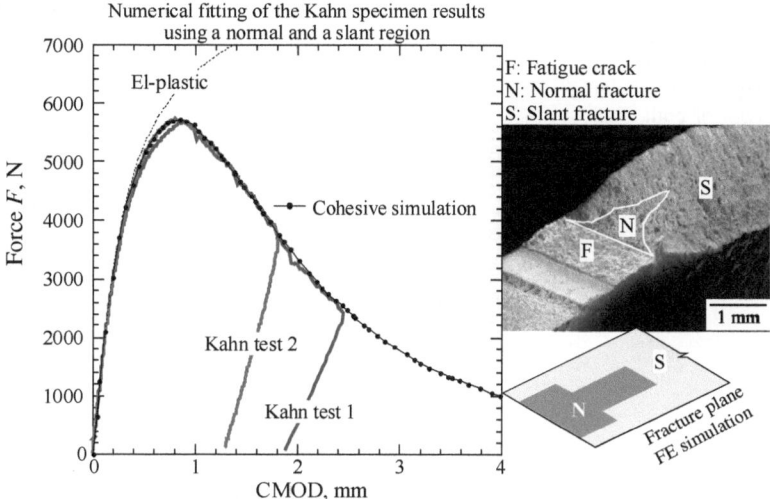

Fig. A.30 Optimization of the cohesive parameters from Fig. A.29, after [7]. Also shown is the loading curve for elastic-plastic behaviour

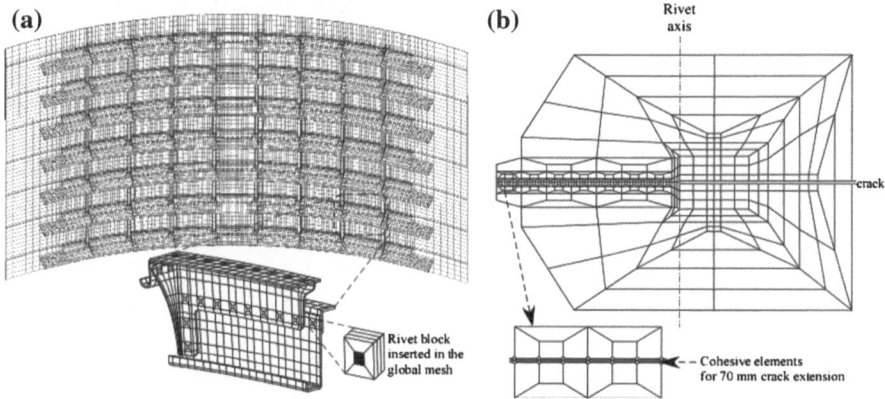

Fig. A.31 Finite element model of the panel; **a** whole structure, with frames assembled from unit cells and rivets represented as blocks; **b** details at the crack tip; after [7]

A.6.3 Finite Element Model

The panel was represented by a 3D CAD model. The model consisted of 77.208 solid elements with 551.086 degrees of freedom and 20 nodes and eight integration points per element. At both crack tips 702 cohesive elements were inserted with 0.2 mm length in crack extension direction and a width equal to the material thickness. These elements with 16 nodes covered a length ahead of the crack tip of 70 mm (Fig. A.31).

Fig. A.32 Simulated internal pressure versus crack extension behaviour of the panel for loading case A, showing the behavior of both crack tips, situated near frames C3 and C5, respectively, after [7]

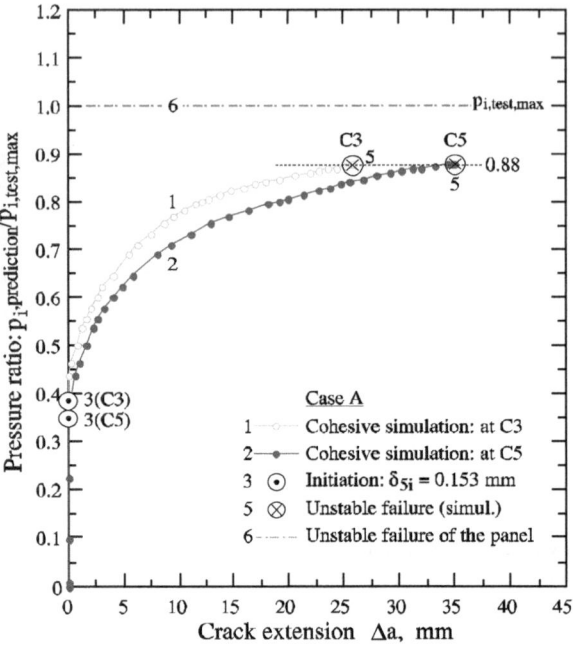

Fig. A.33 Comparison of the simulated crack at frame C3 for cases A and B, after [7]

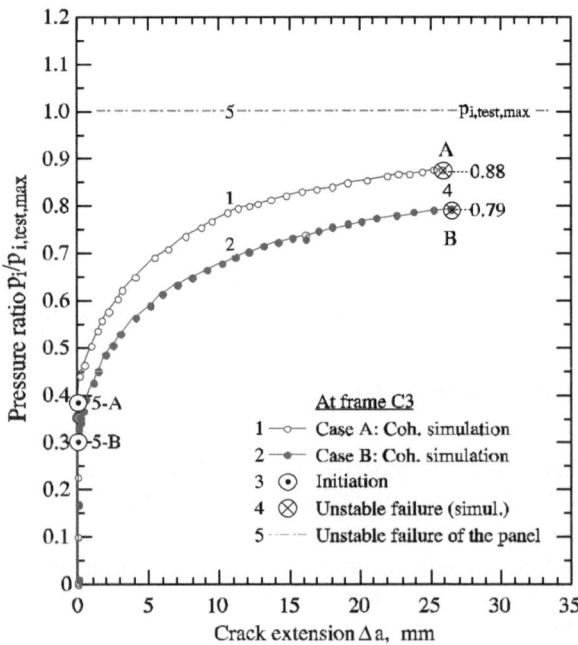

(a)   **(b)**

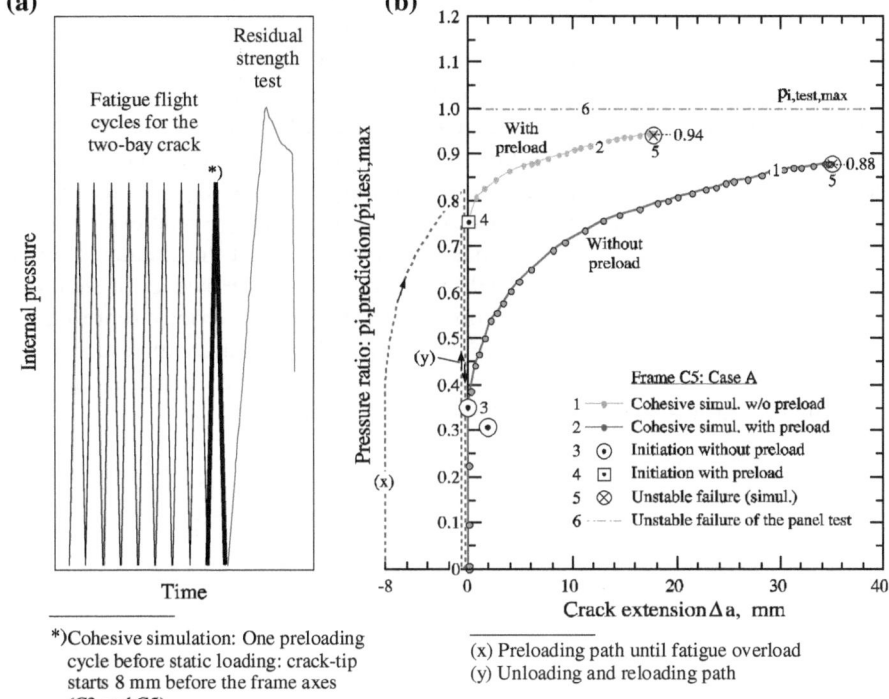

*)Cohesive simulation: One preloading cycle before static loading: crack-tip starts 8 mm before the frame axes (C3 and C5).

(x) Preloading path until fatigue overload
(y) Unloading and reloading path

Fig. A.34 Simulation of the panel test accounting for the cyclic pre-load, after [7]; **a** experimental conditions during cyclic pressure and subsequent fracture test; **b** simulations with and without one cycle of pre-load. The letters x and y refer to corresponding situations in test and simulation

Although the panel components other than the skin had material properties which were substantially different from those of the skin, the whole model was given the stress–strain properties of the skin since the stresses in the stiffeners did not affect the deformation of the whole panel. Two loading conditions of the panel model, A and B, were analysed. Case B modelled the test conditions, whereas case A was supposed to represent better the conditions of the panel as a part of the real fuselage; for details see the original work [7].

A.6.4 Results

Estimation of the residual strength of the panel requires the determination of crack extension, Δa, as a function of the internal pressure, p_i. Therefore, the complete curves p_i versus Δa curve for both crack tips of case B are depicted in Fig. A.32. Both crack tips behave somewhat differently due to the slight asymmetry of the crack with respect to the frames.

A comparison of both loading cases is depicted in Fig. A.33, showing the behaviour of only one crack tip. The simulation that represents the actual test conditions is more conservative than that for case A.

After these simulations, an inquiry at the test lab revealed that the fatigue load used to extend the saw cut to the required final crack length was extremely high, almost 90 % of the failure load in the following test. This was due to the very low loading frequency that could be achieved during pre-cracking. Hence, a new simulation run was undertaken for case A, with a single fatigue load cycle prior to the fracture test simulation. It is obvious from Fig. A.34 that the simulated failure pressure has now been augmented to 94 % of the test load. However, it should be borne in mind that the loading conditions of case A do not strictly represent the test, a case A simulation would still result in a lower failure pressure value. According to Fig. A.33, the failure pressure can be expected to be about 10 % below that of case A, hence about 0.84 times the test value.

Appendix B
Hints for the Treatment of Brittle Materials

For brittle crack extension, a direct identification method of the TSL based on uniaxial tensile test was already proposed in 1981 [8]. However, nowadays it is commonly agreed that only the cohesive strength can be taken from maximum force in a tensile test. The complete TSL should then be identified by inverse methods using a prescribed shape of the TSL, see e.g. [9, 10]. This shape is in general characterised by a separation that starts at maximum cohesive stress followed by a descending section of the function. In the literature, this behaviour is usually approximated by one of the following three functions

- Linearly decreasing, Fig. B.1a
- Bilinearly decreasing, Fig. B.1b
- Exponentially decreasing, Fig. B.1c.

The exponentially decreasing function can be written as

$$T = T_0 \left(1 - \frac{1 - \exp(C\delta/\delta_0)}{1 - \exp(C)} \right) \tag{B.1}$$

with T_0, δ_0 and C being model parameters. Please note that in the concrete fracture community, the cohesive strength, T_0, is usually denoted as tensile strength, f_t, and the material separation is named w with w_c being its critical value.

A further example is shown in Fig. B.1d where both linearly increasing and decreasing sections form the TSL. The understanding of this is that the cohesive layer behaves linearly elastic (stiffness E) until damage initiation. Of course the linear elastic behaviour in the beginning can be combined with all shapes shown in Figs. B.1a–c. In commercial finite element codes, where an initially infinite stiffness is usually not possible, the linear elastic behaviour can precede any separation behaviour.

A common procedure for the identification of the cohesive strength is to use a tensile bar that is pulled to fracture. The value of T_0 is determined by the fracture stress, which is the force divided by the actual area, A, of the specimen at fracture,

K.-H. Schwalbe et al., *Guidelines for Applying Cohesive Models to the Damage Behaviour of Engineering Materials and Structures*, SpringerBriefs in Applied Sciences and Technology, DOI: 10.1007/978-3-642-29494-5, © The Author(s) 2013

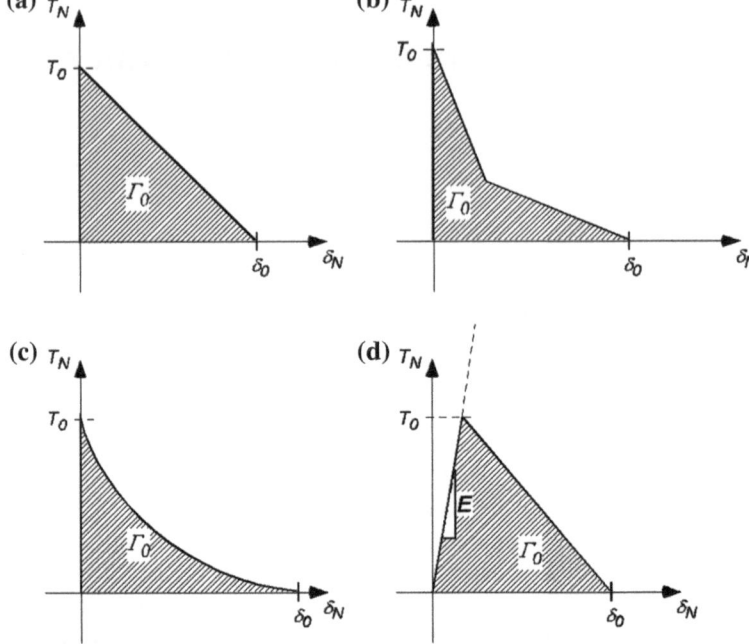

Fig. B.1 Various shapes of traction-separation laws for modelling failure of brittle materials

$T_0 = F/A$. Another procedure is the so-called Brazilian disc or split-cylinder test, in which a cylindrical specimen is loaded under compression.

Under ideally brittle conditions, the cohesive energy, Γ_0 is equal to the linear elastic strain energy release rate, G, which equals twice the materials surface energy, 2γ. Therefore, if a K_{Ic} value is available for the material under consideration, G and thus Γ_0 can be calculated from

$$\Gamma_0 = G_c = \left(K_{Ic}\right)^2 \Big/ E'. \tag{B.2}$$

If other TSLs rather than the simple linear softening law are used, then the additional shape parameters should be fitted numerically to experiments.

References

1. Lin, G., Cornec, A.: Numerische Untersuchung zum Verhalten von Risswiderstandskurven: Simulation mit dem Kohäsivmodell. Materialwiss Werkstofftech **27**, 252–258 (1996)
2. Lin, G., Cornec, A., Schwalbe, K.-H.: Three-dimensional finite element simulation of crack extension in aluminium alloy 2024FC. Fatigue Fract. Eng. Mater. Struct. **21**, 1159–1173 (1998)
3. Cornec, A., Scheider, I., Schwalbe, K.-H.: On the practical application of the cohesive model. Eng. Fract. Mech. **70**, 1963–1987 (2003)

4. ISO IS 12135 Metallic materials—unified method of test for the determination of quasistatic fracture toughness, International Organization for Standardization, Geneva, 2007
5. Yuan, H., Cornec, A., Schwalbe, K.-H.: Numerische Simulation duktilen Risswachstums mittels Kohäsivzonenmodell an dünnen Aluminium CT-Proben. In: Proceedings 23th Session of *DVM-Arbeitskreis "Bruchvorgänge"*, Berlin, Deutscher Verband für Materialprüfung, Berlin, 1991
6. Scheider, I., Brocks, W.: Residual strength prediction of a complex structure using crack extension analyses. Eng. Fract. Mech. **76**, 149–163 (2009)
7. Cornec, A., Schönfeld, W., Schwalbe, K.-H., Scheider, I.: Application of the cohesive model for predicting the residual strength of a large scale fuselage structure with a two-bay crack. Eng. Fail. Anal. **16**, 2541–2558 (2009)
8. Petersson, P.E.: Crack growth and development of fracture zones in plain concrete and similar materials. Report TVBM-1006, Division of Building Materials, Lund Institute of Technology, Lund, Sweden, 1981
9. Elices, M., Planas, J.: Fracture mechanics parameters of concrete. An overview. Adv. Cem. Based Mater. **4**, 116–127 (1996)
10. Guo, X.H., Tin-Loi, F., Li, H.: Determination of quasibrittle fracture law for cohesive crack models. Cem. Concr. Res. **29**, 1055–1059 (1999)